I0055893

Applied
Graph Theory

An Introduction with Graph Optimization and Algebraic Graph Theory

Applied Graph Theory

An Introduction with Graph Optimization and Algebraic Graph Theory

Christopher Griffin

Pennsylvania State University, USA

NEW JERSEY • LONDON • SINGAPORE • BEIJING • SHANGHAI • HONG KONG • TAIPEI • CHENNAI • TOKYO

Published by

World Scientific Publishing Co. Pte. Ltd.
5 Toh Tuck Link, Singapore 596224
USA office: 27 Warren Street, Suite 401-402, Hackensack, NJ 07601
UK office: 57 Shelton Street, Covent Garden, London WC2H 9HE

Library of Congress Control Numbr: 2023015062

British Library Cataloguing-in-Publication Data
A catalogue record for this book is available from the British Library.

APPLIED GRAPH THEORY
An Introduction with Graph Optimization and Algebraic Graph Theory

ISBN 978-981-127-310-0 (hardcover)
ISBN 978-981-127-311-7 (ebook for institutions)
ISBN 978-981-127-312-4 (ebook for individuals)

For any available supplementary material, please visit
https://www.worldscientific.com/worldscibooks/10.1142/13327#t=suppl

Desk Editors: Sanjay Varadharajan/Ana Ovey

Typeset by Stallion Press
Email: enquiries@stallionpress.com

This book is dedicated to Amy and Finn

About the Author

Christopher Griffin is a research professor at the Applied Research Laboratory (ARL), where he holds a courtesy appointment as Professor of Mathematics at Penn State. He was a Eugene Wigner Fellow in the Computational Science and Engineering Division of the Oak Ridge National Laboratory and has also taught in the Mathematics Department at the United States Naval Academy. His favorite part of teaching is connecting complex mathematical concepts with their applications. He finds most students are excited by math when they know how much of our modern world depends on it.

When he is not teaching (which is most of the time), Dr. Griffin's research interests are in applied mathematics, where he focuses on applied dynamical systems (especially on graphs), game theory, and optimization. His research has been funded by the National Science Foundation, the Office of Naval Research, the Army Research Office, the Intelligence Advanced Research Projects Agency, and the Defense Advanced Research Projects Agency. He has published over 100 peer-reviewed research papers in various fields of applied mathematics.

Preface

Why did I write this book? This book started in 2011 as a set of lecture notes called *Graph Theory: Penn State Math 485 Lecture Notes*. I wrote a portion of those notes while staying in Key West, Florida. I highly recommend writing while in Key West. Actually, I recommend doing just about anything in Key West (or some other tropical place) if you can get there.

Until 18 months ago, I was perfectly content to ignore my colleagues when they told me that I should turn these notes into a book. Nevertheless, here we are, and I should tell you something about this book and why it's different from all the other graph theory books. To understand that, it's best to understand why I wrote the lecture notes in the first place.

Math 485 is Penn State Math's advanced undergraduate course on graph theory. It is taken by students after they have passed a course in discrete mathematics with a focus on proof. I've now taught this course several times. Originally, I wrote the lecture notes because, when I was young, I lived in terror of getting lost in the middle of a lecture and wasting 10 minutes doing the "absent-minded professor thing." I don't worry about that as much anymore. Key West can help with things like this as well.

The first few times I taught the course, I included a module on linear programming and covered network flows using the formalism of the Karush–Kuhn–Tucker conditions rather than using a traditional approach. I rather liked this, and it helped introduce the students to connections across mathematics. I am a fan of algebraic graph theory, so we covered material on graphs and matrices prior to jumping into linear programming.

While I was at the United States Naval Academy, I taught SA403: Graph and Network Algorithms. This class needed to cover classic graph algorithms, so I changed up the notes to cover classic network flow. I also taught a class called Intermediate Linear Algebra, which had an emphasis on applications of linear algebra. This was the perfect place to use topics from algebraic graph theory, so I wrote some new material on the Lagrangian matrix and spectral clustering.

Since I usually teach undergraduates, and most (but not all) undergraduates like to see some applications, I also tried to cover as many applications as possible without disrupting the flow of a "theorem-proof"-type class. Letting the theory drive the applications suits me perfectly because I'm a very applied mathematician.

Eighteen months ago, a very nice editor named Rochelle from World Scientific contacted me and asked if I'd be willing to turn my notes into a book. I had been asked this before and always said "no" because I really didn't want the hassle. Rochelle promised faithfully that she would leave me alone and let me set the timeline, and, amazingly, that's what happened. The result is this book, which tries to combine the best of all the things I liked about my lecture notes into something other people might be able to use.

How do I use this book? This book can be used entirely for self-study or in a classroom. It is really designed for a one-semester course and is geared toward undergraduates who have taken a class on proof techniques (usually called Discrete Mathematics). However, clever undergraduates can probably read the material with very little problem even if they haven't taken a course on proofs. The book is organized into four parts and written in a "theorem-proof-example" style. There are remarks throughout as well as chapter notes that try to emphasize some history or other applications. Part 1 covers introductory material on graphs that would be standard in any graph theory class. Part 2 covers algorithms, network flows, and coloring. Part 3 covers algebraic graph theory. Finally, Part 4 covers linear programming and network flow problems, including an alternate proof for the max-flow/min-cut theorem using the Karush–Kuhn–Tucker conditions. Each part emphasizes some kind of application, with vertex ranking (centrality) appearing throughout the book. The coloring section has a nice proof of the NP-completeness of k-colorability and demonstrates the connection between mathematical logic and combinatorics.

Sample curriculum is shown in the following figure.

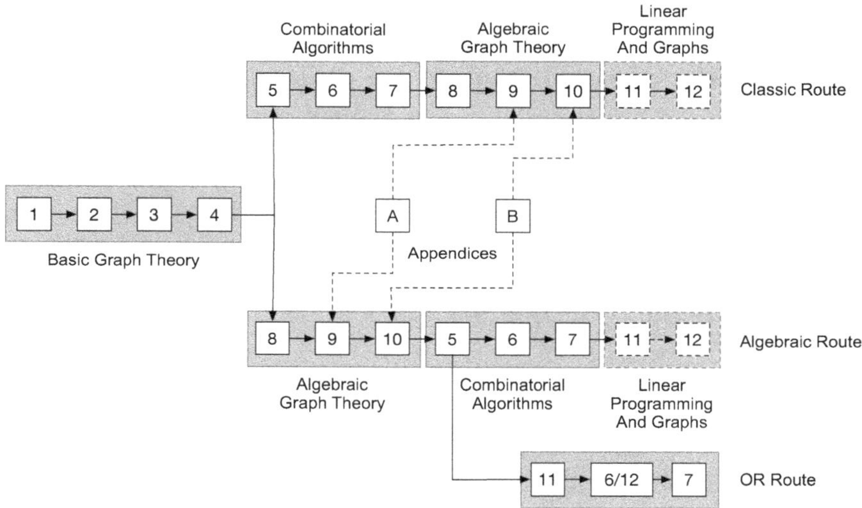

I've used each possible path through the curriculum at least once and they all worked for me.

Classic route: The classic route starts with an introduction to graph theory (and some applications), goes through graph algorithms (Prim's, Dijkstra's, etc.), handles coloring and NP-completeness, and then covers algebraic graph theory. The material on linear programming is optional or could be used for extra credit. You can choose to take up coloring later, which is how I did it at the naval academy.

Algebraic route: The algebraic route starts with an introduction to graph theory and then immediately transitions to algebraic graph theory. This introduces students more quickly to the connections between graph theory and other areas of mathematics. This also allows you to cover all aspects of "centrality" more or less in sequence. This path then circles back to cover other graph algorithms. Again, the material on linear programming is optional.

Operations research route: This route follows the algebraic route but then deviates to cover linear programming immediately after covering classic graph algorithms. Using material from both Chapters 6 and 12 together, the max-flow/min-cut theorem is covered through the lens of linear programming, and this is used to derive the Edmonds–Karp algorithm. This is how I taught Math 485 the first few times.

The book is sensitive to the fact that linear programming can be a niche interest in some math departments. Consequently, the coverage of network flows is classical in Chapter 6, which makes it easier for instructors who want to follow the classical route. For the more adventurous, combining Chapters 6 and 12 (which does require a little back-and-forth) does provide a unique perspective on network flows that is not usually available at the undergraduate level. Additionally, there are two appendices on linear algebra and probability to help students who have not had sufficient coverage of these topics. This is relevant when dealing with algebraic graph theory.

My favorite aspect of the book is the applications. Degree, betweenness, eigenvector, and page rank centrality are all covered, as are spectral clustering and the graph Laplacian. The graph algorithms are easy to understand with applications to routing, but I also discuss arbitrage discovery in currency exchanges (which was a question I was asked during an Amazon interview). Computing whether sports teams can make the playoffs is discussed as an application of network flows, and exam scheduling illustrates the use of coloring. For those who cover the linear programming material, the assignment problem is discussed along with profit maximization in simple companies.

What isn't in this book? Since this is geared toward a one-semester class, there are several omissions. For those who wish to try the linear programming material, the simplex algorithm is not presented. Instead, I show how to solve linear programs with a computer. (However, I do cover optimality conditions and duality.) Planar graphs are entirely ignored in the main text, though they are briefly discussed (along with the four color theorem) in chapter notes. If you've found my lecture notes online, you'll notice I have expunged coverage of random graphs in favor of the graph Laplacian. This was a difficult decision, but (i) the chapter did not fit into the flow and

(ii) it was too similar to the coverage by Gross and Yellen in their second edition of *Graph Theory and its Applications*. While I seriously considered including coverage of edge and vertex spaces (including cut and cycle spaces), including sufficient background on direct sums in linear algebra would have made the book too long to pass as a one-semester treatment. Also, I never managed to cover those in a class, so I don't know how students would react.

Acknowledgments

This book was created with LaTeX2e using TeXShop and Bibdesk. Figures were created using Mathematica™ and Omnigraffle™. In acknowledging those people who helped make this work possible, first let me say thank you to all the other scholars who have written texts on graph theory. Without your pioneering works, there would be nothing to write about. Also, I must thank the individuals who found typos in my original lecture notes and wrote to tell me about them: Suraj Shekhar and Prof. Elena Kosygina. I would very much like to thank my colleagues at Penn State for encouraging me to turn my notes into a book. Finally, I owe a debt of gratitude to Rochelle Kronzek Miller of World Scientific, who gave me the final push to write this, and to Ana Ovey and Sanjay Varadharajan, who dealt with the mechanics of getting this published. Thank you all.

Contents

Part 1

Introduction to Graphs

Chapter 1

Introduction to Graph Theory

Remark 1.1 (Chapter goals). The goal of this chapter is to introduce the basic concepts of graph theory and its vocabulary. We also discuss a little of its history.

1.1 Graphs, Multigraphs, and Simple Graphs

Definition 1.2 (Graph). A graph is a tuple $G = (V, E)$ where V is a (finite) set of vertices and E is a finite collection of edges. The set E contains elements from the union of the one- and two-element subsets of V. That is, each edge is either a one- or two-element subset of V.

Remark 1.3. It is generally easiest to visualize a graph as a collection of shapes for its vertices and lines (or curves) connecting the vertices for its edges. Though any shape can be used for the vertices, in mathematical graph theory, dots or circles are the most common.

Example 1.4. Consider the set of vertices $V = \{1, 2, 3, 4\}$ and the set of edges

$$E = \{\{1, 2\}, \{2, 3\}, \{3, 4\}, \{4, 1\}\}.$$

Then, the graph $G = (V, E)$ has four vertices and four edges. See Fig. 1.1 for the visual representation of G.

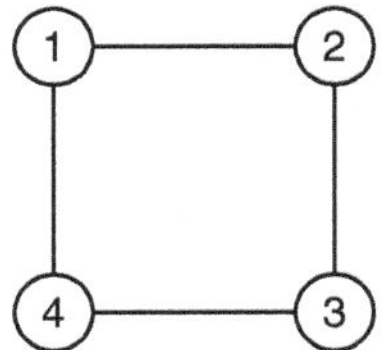

Fig. 1.1 It is easier for explanations to represent a graph by a diagram in which vertices are represented by points (or squares, circles, triangles, etc.) and edges are represented by lines connecting vertices.

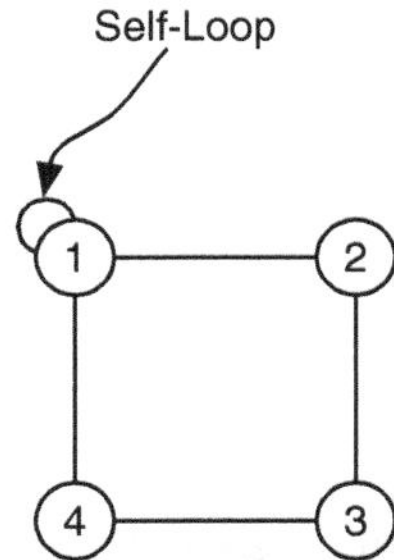

Fig. 1.2 A self-loop is an edge in a graph G that contains exactly one vertex. That is, an edge that is a one-element subset of the vertex set. Self-loops are illustrated by loops at the vertex in question.

Definition 1.5 (Self-loop). If $G = (V, E)$ is a graph and $v \in V$ and $e = \{v\}$, then edge e is called a *self-loop*. That is, any edge that is a single-element subset of V is called a self-loop.

Example 1.6. If we replace the edge set in Example 1.4 with

$$E = \{\{1, 2\}, \{2, 3\}, \{3, 4\}, \{4, 1\}, \{1\}\},$$

then the visual representation of the graph includes a self-loop that starts and ends at Vertex 1. This is illustrated in Fig. 1.2.

Definition 1.7 (Vertex adjacency). Let $G = (V, E)$ be a graph. Two vertices v_1 and v_2 are said to be *adjacent* if there exists an edge $e \in E$ so that $e = \{v_1, v_2\}$. A vertex v is self-adjacent if $e = \{v\}$ is an element of E.

Definition 1.8 (Edge adjacency). Let $G = (V, E)$ be a graph. Two edges e_1 and e_2 are said to be *adjacent* if there exists a vertex

v so that v is an element of both e_1 and e_2 (as sets). An edge e is said to be *adjacent* to a vertex v if v is an element of e as a set.

Definition 1.9 (Neighborhood). Let $G = (V, E)$ be a graph, and let $v \in V$. The *neighbors* of v are the set of vertices that are adjacent to v. Formally,

$$N(v) = \{u \in V : \exists e \in E\,(e = \{u, v\} \text{ or } u = v \text{ and } e = \{v\})\}. \quad (1.1)$$

In some texts, $N(v)$ is called the *open neighborhood* of v, while $N[v] = N(v) \cup \{v\}$ is called the *closed neighborhood* of v. This notation is somewhat rare in practice. When v is an element of more than one graph, we write $N_G(v)$ as the neighborhood of v in graph G.

Remark 1.10. Equation (1.1) is read:

> $N(v)$ is the set of vertices u in (the set) V such that there exists an edge e in (the set) E so that $e = \{u, v\}$ or $u = v$ and $e = \{v\}$.

The logical expression $\exists x\,(R(x))$ is always read in this way; that is, there exists x so that some statement $R(x)$ holds. Similarly, the logical expression $\forall y\,(R(y))$ is read:

> For all y the statement $R(y)$ holds.

Admittedly, this sort of thing is very pedantic, but logical notation can help immensely in simplifying complex mathematical expressions.

Remark 1.11. The difference between the open and closed neighborhoods of a vertex can get a bit odd when you have a graph with self-loops. Since this is a highly specialized case, usually the author (of the paper, book, etc.) will specify a behavior.

Example 1.12. In the graph in Example 1.12, the neighborhood of Vertex 1 consists of Vertices 2 and 4 and Vertex 1 because Vertex 1 is adjacent to itself. The neighborhood of Vertex 1 now consists of vertices 1, 2, and 4. Note that the self-loop forces Vertex 1 into its own neighborhood.

Definition 1.13 (Degree). Let $G = (V, E)$ be a graph, and let $v \in V$. The *degree* of v, written $\deg(v)$, is the number of non-self-loop edges adjacent to v plus two times the number of self-loops defined at v. More formally,

$$\deg(v) = |\{e \in E : \exists u \in V(e = \{u, v\})\}| + 2\,|\{e \in E : e = \{v\}\}|\,.$$

Here, if S is a set, then $|S|$ is the cardinality of that set.

Remark 1.14. Note that each vertex in the graph in Fig. 1.1 has a degree of 2.

Example 1.15. Consider the graph shown in Example 1.6. The degree of Vertex 1 is 4. We obtain this by counting the number of non-self-loop edges adjacent to Vertex 1 (there are 2) and adding two times the number of self-loops at Vertex 1 (there is 1) to obtain $2 + 2 \times 1 = 4$.

Remark 1.16. Let $G = (V, E)$ be a graph. There are two degree values that are of interest in graph theory: The largest and smallest vertex degrees are usually denoted by $\Delta(G)$ and $\delta(G)$, respectively. That is,

$$\Delta(G) = \max_{v \in V} \deg(v) \text{ and} \tag{1.2}$$

$$\delta(G) = \min_{v \in V} \deg(v). \tag{1.3}$$

Definition 1.17 (Multigraph). A graph $G = (V, E)$ is a *multigraph* if there are two edges e_1 and e_2 in E so that e_1 and e_2 are equal as sets. That is, there are two vertices v_1 and v_2 in V so that $e_1 = e_2 = \{v_1, v_2\}$.

Remark 1.18. Note in the definition of graph (Definition 1.2), we were very careful to specify that E is a *collection* of one- and two-element subsets of V rather than to say that E was a set. This allows us to have duplicate edges in the edge set and thus to define multigraphs. In computer science, a set that may have duplicate entries is sometimes called a *multiset*. A multigraph is a graph in which E is a multiset.

Derivation 1.19 (Königsburg bridge problem). Graph theory began with Leonhard Euler with his study of the bridges of Königsburg problem. Here's how it started: The city of Königsburg exists as a collection of islands connected by bridges. The problem

Euler wanted to analyze was: Is it possible to go from island to island, traversing each bridge only once?

Following Euler, we construct a graph to analyze the bridges of Königsburg problem. Assume that we treat each island as a vertex and each bridge as an edge. The resulting multigraph is illustrated in Fig. 1.3. The edge collection is

$$E = \{\{A, B\}, \{A, B\}, \{A, C\}, \{A, C\}, \{A, D\}, \{B, D\}, \{C, D\}\}.$$

This multigraph occurs because there are two bridges connecting island A with island B and two bridges connecting island A with island C. If two vertices are connected by two (or more) edges, then the edges are simply represented as parallel lines (or arcs) connecting the vertices.

Note that this representation dramatically simplifies the analysis of the problem in so far as we can now focus only on the structural properties of this graph. It's easy to see (from Fig. 1.3) that each vertex has an odd degree. More importantly, since we are trying to traverse islands without ever recrossing the same bridge (edge), when we enter an island (say C), we use one of the three edges. Unless this is our final destination, we must use *another* edge to leave C. Additionally, assuming we have not crossed all the bridges yet, we know that we *must* leave C. That means that the third edge that

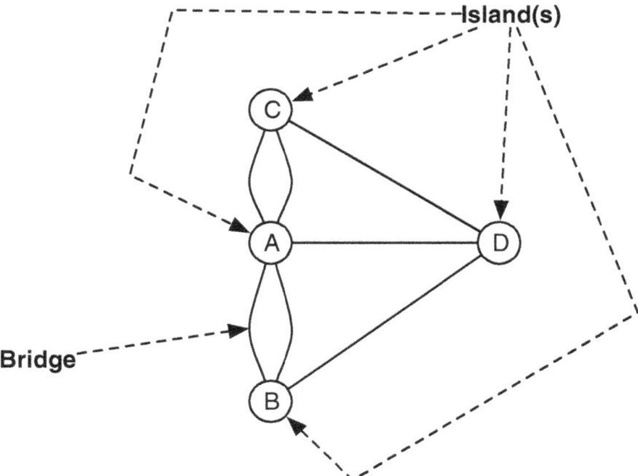

Fig. 1.3 Representing each island as a dot and each bridge as a line or curve connecting the dots simplifies the visual representation of the seven Königsburg bridges.

touches C *must* be used to return to C a final time. Alternatively, we could *start* at Island C and then return once and never come back. Put simply, our trip around the bridges of Königsburg had better start *or* end at Island C. But Islands (vertices) B and D *also* have this property. We can't start and end our travels over the bridges on Islands C, B, and D simultaneously; therefore, no such walk around the islands in which we cross each bridge precisely once is possible.

Definition 1.20 (Simple graph). A graph $G = (V, E)$ is a simple graph if G has no edges that are self-loops and if E is a *subset* of two-element subsets of V, i.e., G is not a multigraph.

Remark 1.21. Most of graph theory is not concerned with graphs containing either self-loops *or* multigraphs. Thus, we assume that every graph we discuss from this point on is a *simple* graph, and we use the term *graph* to mean *simple graph.* When a particular result holds in a more general setting, we state it explicitly.

1.2 Directed Graphs

Definition 1.22 (Directed graph). A directed graph (digraph) is a tuple $G = (V, E)$ where V is a (finite) set of vertices and E is a collection of elements contained in $V \times V$. That is, E is a collection of ordered pairs of vertices. The edges in E are called *directed edges* to distinguish them from those edges in Definition 1.2

Definition 1.23 (Source/Destination). Let $G = (V, E)$ be a directed graph. The *source* (or *tail*) of the (directed) edge $e = (v_1, v_2)$ is v_1, while the *destination* (or *sink* or *head*) of the edge is v_2.

Remark 1.24. A directed graph (digraph) differs from a graph only insofar as we replace the concept of an edge as a set with the idea that an edge is an ordered pair in which the ordering gives some notion of the direction of a flow. In the context of a digraph, a *self-loop* is an ordered pair with form (v, v). We can define a multi-digraph if we allow the set E to be a true collection (rather than a set) that contains multiple copies of an ordered pair.

Remark 1.25. It is worth noting that the ordered pair (v_1, v_2) is *distinct* from the pair (v_2, v_1). Thus, if a digraph $G = (V, E)$ has both (v_1, v_2) and (v_2, v_1) in its edge set, it is *not* a multi-digraph.

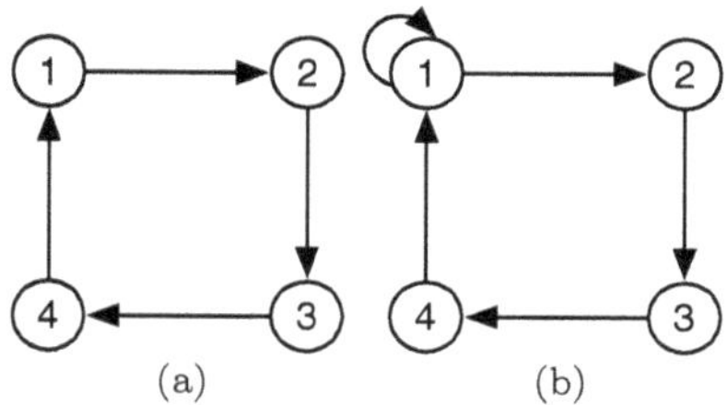

Fig. 1.4 (a) A directed graph and (b) a directed graph with a self-loop. In a directed graph, edges are directed; that is, they are ordered pairs of elements drawn from the vertex set. The ordering of the pair gives the direction of the edge.

Example 1.26. We can modify the figures in Example 1.4 to make them directed. Suppose we have the directed graph with vertex set $V = \{1, 2, 3, 4\}$ and edge set

$$E = \{(1, 2), (2, 3), (3, 4), (4, 1)\}.$$

This digraph is visualized in Fig. 1.4(a). In drawing a digraph, we simply append arrowheads to the destination associated with a directed edge.

We can likewise modify our self-loop example to make it directed. In this case, our edge set becomes

$$E = \{(1, 2), (2, 3), (3, 4), (4, 1), (1, 1)\}.$$

This is shown in Fig. 1.4(b).

Definition 1.27 (In-degree, out-degree). Let $G = (V, E)$ be a digraph. The *in-degree* of a vertex v in G is the total number of edges in E with *destination* v. The *out-degree* of v is the total number of edges in E with *source* v. We denote the in-degree of v by $\deg_{\text{in}}(v)$ and the out-degree by $\deg_{\text{out}}(v)$.

Definition 1.28 (Underlying graph). If $G = (V, E)$, is a digraph, then the underlying graph of G is a (multi)graph (with self-loops) that results when each directed edge (v_1, v_2) is replaced by the set $\{v_1, v_2\}$, thus making the edge nondirectional. Naturally, if the directed edge is a directed self-loop (v, v), then it is replaced by the singleton set $\{v\}$.

Remark 1.29. Notions like edge and vertex adjacency and neighborhood can be extended to digraphs by simply defining them with respect to the underlying graph of a digraph. Thus, the neighborhood of a vertex v in a digraph G is $N(v)$ computed in the underlying graph.

Remark 1.30. Whether the underlying graph of a digraph is a multigraph or not usually has no bearing on relevant properties. In general, an author will state whether two directed edges (v_1, v_2) and (v_2, v_1) are combined into a single set $\{v_1, v_2\}$ or two sets in a multiset. As a rule of thumb, multi-digraphs will have underlying multigraphs, while digraphs generally have underlying graphs that are not multigraphs.

Remark 1.31. It is possible to mix (undirected) edges and directed edges together into a very general definition of a graph with both undirected and directed edges. This usually only occurs in specific models, and we will not consider such graphs. For the remainder of this book, unless otherwise stated:

(1) When we say *graph*, we mean a *simple graph*, as in Remark 1.21.
(2) When we say *digraph*, we mean a directed graph $G = (V, E)$, in which every edge is a directed edge and the component E is a *set*. For practical purposes (as in when we discuss Markov chains), we allow self-loops in digraphs.

1.3 Chapter Notes

Leonhard Euler (1707–1783), the father of graph theory, was born in Basel, Switzerland, and studied under the Bernoullis. He is known as one of the greatest mathematicians of all time. He is one of the most prolific mathematical writers in history, with over 850 papers, some published after his death [1]. Euler conceived of the basic elements of graph theory after being presented with the bridges of Königsburg problem, as detailed in this letter to Giovanni Jacopo Marinoni [2]:

> *A problem was posed to me about an island in the city of Königsberg, surrounded by a river spanned by seven bridges, and I was asked whether someone could traverse the separate bridges in a connected walk in such a way that each bridge is crossed only once. I was informed that hitherto no-one had*

demonstrated the possibility of doing this, or shown that it is impossible. This question is so banal, but seemed to me worthy of attention in that geometry, nor algebra, nor even the art of counting was sufficient to solve it. In view of this, it occurred to me to wonder whether it belonged to the geometry of position [geometriam Situs], which Leibniz had once so much longed for. And so, after some deliberation, I obtained a simple, yet completely established, rule with whose help one can immediately decide for all examples of this kind, with any number of bridges in any arrangement, whether such a round trip is possible, or not...

Leonard Euler
Letter to Giovanni Jacopo Marinoni
1736

1.4 Exercises

Exercise 1.1
In an online social network, Alice is friends with Bob and Charlie. Charlie is friends with David and Edward. Edward is friends with Bob. Draw a graph to represent this social network.

Exercise 1.2
Find the neighborhoods and degrees for each vertex in the graph shown in Fig. 1.5.

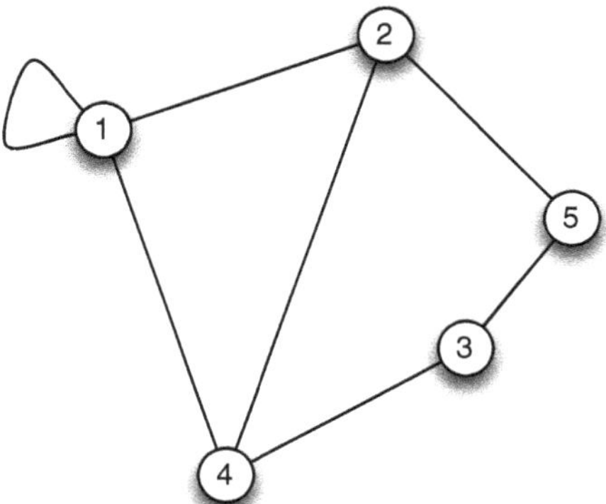

Fig. 1.5 Graph for Exercise 1.2.

Exercise 1.3

Since Euler's work, two of the seven bridges in Königsburg have been destroyed (during World War II). Another two were replaced by major highways, but they are still (for all intents and purposes) bridges. The remaining three are still intact (see Fig. 1.6). Determine whether it is possible to visit the bridges traversing each bridge exactly once. If so, find such a sequence of edges.

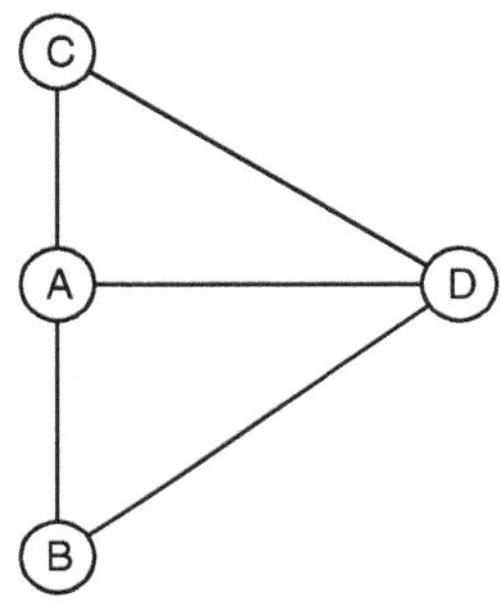

Fig. 1.6 During World War II, two of the seven original Königsburg bridges were destroyed. Later, two more were made into modern highways (but they are still bridges). Is it now possible to go from island to island, traversing each bridge only once?

Exercise 1.4

Consider the new bridges of Königsburg problem from Exercise 1.3. Is the graph representation of this problem a simple graph? Could a self-loop exist in a graph derived from a bridges-of-Königsburg-type problem? If so, what would it mean? If not, why?

Exercise 1.5

Prove that for simple graphs, the degree of a vertex is simply the cardinality of its (open) neighborhood.

Exercise 1.6

Suppose in the new bridges of Königsburg problem (from Exercise 1.3), some of the bridges are to become *one way*. Find a way of replacing the edges in the graph you obtained in solving Exercise 1.3 with directed edges so that the graph becomes a digraph, but it is still possible to tour all the islands without crossing the same bridge twice. Is it possible to directionalize the edges so that a tour in which each bridge is crossed once is not possible, but it is still possible to enter and exit each island? If so, do it. If not, prove that it is not possible. [Hint: In this case, enumeration is not that hard and is the most straightforward. You can use symmetry to shorten your argument substantially.]

Chapter 2

Degree Sequences and Subgraphs

Remark 2.1 (Chapter goals). In this chapter, we introduce the idea of a degree sequence. We then discuss graph families with special degree sequences, prove the Havel–Hakimi theorem, and discuss subgraphs. We conclude by discussing cliques, independent sets, and vertex covers.

2.1 Degree Sequences

Definition 2.2 (Degree sequence). Let $G = (V, E)$ be a graph, with $|V| = n$. The *degree sequence* of G is a tuple $\mathbf{d} \in \mathbb{Z}^n$ composed of the degrees of the vertices in V arranged in *decreasing order*.

Example 2.3. Consider the graph in Fig. 2.1. The degrees for the vertices of this graph are:

(1) $\mathrm{v}_1 = 4$,
(2) $\mathrm{v}_2 = 3$,
(3) $\mathrm{v}_3 = 2$,
(4) $\mathrm{v}_4 = 2$, and
(5) $\mathrm{v}_5 = 1$.

This leads to the degree sequence $\mathbf{d} = (4, 3, 2, 2, 1)$.

Definition 2.4 (Empty and trivial graphs). A graph $G = (V, E)$ in which $V = \emptyset$ is called an *empty* graph (or *null* graph). A graph in which $V = \{v\}$ and $E = \emptyset$ is called a *trivial* graph.

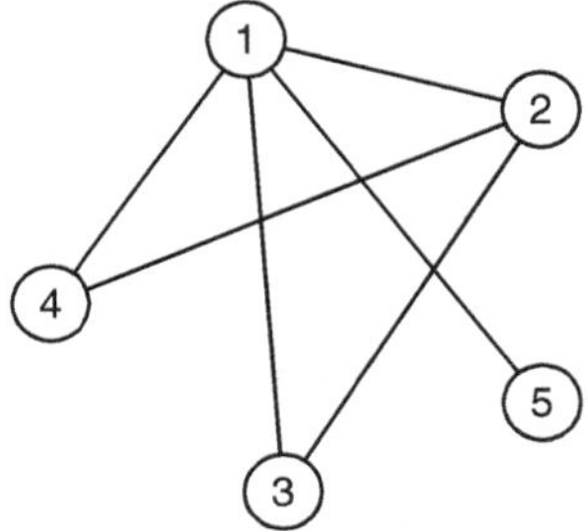

Fig. 2.1 The graph above has a degree sequence $\mathbf{d} = (4, 3, 2, 2, 1)$. These are the degrees of the vertices in the graph arranged in increasing order.

Remark 2.5. An empty graph has an empty degree sequence. A trivial graph has a degree sequence composed of all zeros.

Definition 2.6 (Isolated vertex). Let $G = (V, E)$ be a graph, and let $v \in V$. If $\deg(v) = 0$, then v is said to be *isolated.*

Remark 2.7. Note that Definition 2.6 applies only when G is a simple graph. If G is a general graph (one with self-loops), then v is still isolated even when $\{v\} \in E$; that is, there is a self-loop at vertex v and no other edges are adjacent to v. In this case, however, $\deg(v) = 2$.

Assumption 1 (Pigeonhole principle). Suppose that items may be classified according to m possible types, and we are given $n > m$ items. Then, there are at least two items with the same type.

Remark 2.8. The pigeonhole principle was originally formulated by thinking of placing $m + 1$ pigeons into m pigeon holes. Clearly, to place all the pigeons in the holes, one hole must have two pigeons in it. The holes are the types (each whole is a different type), and the pigeons are the objects. Another good example deals with gloves: There are two types of gloves (left handed and right handed). If I hand you three gloves (the objects), then you either have two left-handed gloves or two right-handed gloves.

Theorem 2.9. *Let $G = (V, E)$ be a nonempty, nontrivial graph. Then, G has at least one pair of vertices with equal degree.*

Proof. This proof uses the pigeonhole principle and is illustrated by the graph in Fig. 2.1, where $\deg(v_3) = \deg(v_4)$. The types will be the possible vertex degree values, and the objects will be the vertices.

Suppose $|V| = n$. Each vertex *could* have a degree between 0 and $n-1$ (for a total of n possible degrees), but if the graph has a vertex of degree 0, then it cannot have a vertex of degree $n-1$. Therefore, there are only at most $n-1$ possible degree values, depending on whether the graph has an isolated vertex or a vertex with degree $n-1$ (if it has neither, there are even fewer than $n-1$ possible degree values). Thus, by the pigeonhole principle, at least two vertices must have the same degree. □

Theorem 2.10. *Let $G = (V, E)$ be a (general) graph, then*

$$2|E| = \sum_{v \in V} \deg(v). \tag{2.1}$$

Proof. Consider two vertices v_1 and v_2 in V. If $e = \{v_1, v_2\}$, then a $+1$ is contributed to $\sum_{v \in V} \deg(v)$ for both v_1 and v_2. Thus, every non-self-loop edge contributes $+2$ to the vertex degree sum. On the other hand, if $e = \{v_1\}$ is a self-loop, then this edge contributes $+2$ to the degree of v_1. Therefore, each edge contributes exactly $+2$ to the vertex degree sum. Equation (2.1) follows immediately. □

Corollary 2.11. *Let $G = (V, E)$. Then, there are an even number of vertices in V with odd degree.* □

Theorem 2.12. *Let $G = (V, E)$ be a digraph. Then, the following holds:*

$$|E| = \sum_{v \in V} \deg_{\text{in}}(v) = \sum_{v \in V} \deg_{\text{out}}(v). \tag{2.2}$$

□

Definition 2.13 (Graphic sequence). Let $\mathbf{d} = (d_1, \ldots, d_n)$ be a tuple in $\mathbb{Z}^n$, with $d_1 \geq d_2 \geq \cdots \geq d_n$. Then, $\mathbf{d}$ is *graphic* if there exists a graph G with degree sequence $\mathbf{d}$.

Remark 2.14. See the chapter notes for a discussion on the applications of graphs with specific kinds of degree distributions.

Corollary 2.15. *If* $\mathbf{d}$ *is graphic, then the sum of its elements is even.* □

Lemma 2.16. *Let* $\mathbf{d} = (d_1, \ldots, d_n)$ *be a graphic degree sequence. Then, there exists a graph* $G = (V, E)$ *with degree sequence* $\mathbf{d}$ *so that if* $V = \{v_1, \ldots, v_n\}$*, then:*

(1) $\deg(v_i) = d_i$ *for* $i = 1, \ldots, n$*; and*
(2) v_1 *is adjacent to vertices* $v_2, \ldots, v_{d_1+1}$*.*

Proof. The fact that $\mathbf{d}$ is graphic means there is at least one graph whose degree sequence is equal to $\mathbf{d}$. From among all those graphs, chose $G = (V, E)$ to maximize

$$r = |N(v_1) \cap \{v_2, \ldots, v_{d_1+1}\}|. \tag{2.3}$$

Recall that $N(v_1)$ is the neighborhood of v_1. Thus, maximizing Eq. (2.3) implies that we are attempting to make sure that as many vertices in the set $\{v_2, \ldots, v_{d_1+1}\}$ are adjacent to v_1 as possible.

If r $= d_1$, then the theorem is proved since v_1 is adjacent to $v_2, \ldots, v_{d_1+1}$. Now, proceed by contradiction and assume $r < d_1$. We know the following things:

(1) Since $\deg(v_1) = d_1$, there must be a vertex v_t with $t > d_1 + 1$ so that v_t is adjacent to v_1.
(2) Moreover, there is a vertex v_s with $2 \leq s \leq d_1 + 1$ that is *not* adjacent to v_1.
(3) By the ordering of V, $\deg(v_s) \geq \deg(v_t)$; that is, $d_s \geq d_t$.
(4) Therefore, there is some vertex $v_k \in V$ so that v_s is adjacent to v_k but v_t is not because v_t is adjacent to v_1 and v_s is not, and the degree of v_s is at least as large as the degree of v_t.

Let us create a new graph $G' = (V, E')$. The edge set E' is constructed from E by:

(1) removing edge $\{v_1, v_t\}$,
(2) removing edge $\{v_s, v_k\}$,
(3) adding edge $\{v_1, v_s\}$, and
(4) adding edge $\{v_t, v_k\}$.

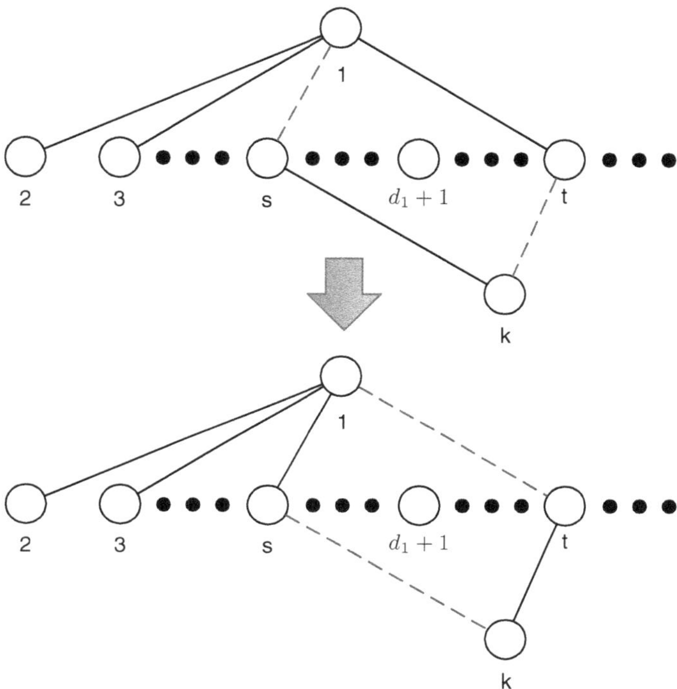

Fig. 2.2 We construct a new graph G' from G that has a larger value r (see Eq. (2.3)) than our original graph G did. This contradicts our assumption that G was chosen to maximize r.

This is illustrated in Fig. 2.2. In this construction, the degrees of v_1, v_t, v_s, and v_k are preserved. However, it is clear that in G',

$$r' = |N_{G'}(v_1) \cap \{v_2, \dots, v_{d_1+1}\}|,$$

and we have $r' > r$. This contradicts our initial choice of G and proves the theorem. □

Theorem 2.17 (Havel–Hakimi theorem). *A degree sequence* $\mathbf{d} = (d_1, \dots, d_n)$ *is graphic if and only if the sequence* $(d_2 - 1, \dots, d_{d_1+1} - 1, d_{d_1+2}, \dots, d_n)$ *is graphic.*

Proof. ($\Rightarrow$) Suppose that $\mathbf{d} = (d_1, \dots, d_n)$ is graphic. Then, by Lemma 2.16, there is a graph G with degree sequence d so that:

(1) $\deg(v_i) = d_i$ for $i = 1, \ldots, n$; and
(2) v_1 is adjacent to vertices $v_2, \ldots, v_{d_1+1}$.

If we remove vertex v_1 and all edges containing v_1 from this graph G to obtain G', then in G', for all $i \in \{2, \ldots d_1 + 1\}$, the degree of v_i is $d_i - 1$, while for $j \in \{d_1 + 2, \ldots, n\}$, the degree of v_j is d_j because v_1 is not adjacent to $v_{d_1+2}, \ldots, v_n$ by choice of G. Thus, G' has a degree sequence of $(d_2 - 1, \ldots, d_{d_1+1} - 1, d_{d_1+2}, \ldots, d_n)$, and thus it is graphic.

($\Leftarrow$) Now, suppose that $(d_2 - 1, \ldots, d_{d_1+1} - 1, d_{d_1+2}, \ldots, d_n)$ is graphic. Then, there is some graph G that has this as its degree sequence. We can construct a new graph G' from G by adding a vertex v_1 to G and creating an edge from v_1 to each vertex from v_2 to v_{d_1+1}. It is clear that the degree of v_1 is d_1, while the degrees of all other vertices v_i must be d_i, and thus, $\mathbf{d} = (d_1, \ldots, d_n)$ is graphic because it is the degree sequence of G'. This completes the proof. □

Remark 2.18. Naturally, one might have to rearrange the ordering of the degree sequence $(d_2 - 1, \ldots, d_{d_1+1} - 1, d_{d_1+2}, \ldots, d_n)$ to ensure it is in descending order.

Example 2.19. Consider the degree sequence $\mathbf{d} = (5, 5, 4, 3, 2, 1)$. One might ask if this degree sequence is graphic. Note that $5 + 5 + 4 + 3 + 2 + 1 = 20$, so, at least, the necessary condition that the degree sequence sum to an even number is satisfied. In this $\mathbf{d}$, we have $d_1 = 5$, $d_2 = 5$, $d_3 = 4$, $d_4 = 3$, $d_5 = 2$, and $d_6 = 1$.

Applying the Havel–Hakimi theorem, we know that this degree sequence is graphic if and only if $\mathbf{d}' = (4, 3, 2, 1, 0)$ is graphic. Note that this is $(d_2 - 1, d_3 - 1, d_4 - 1, d_5 - 1, d_6 - 1)$ since $d_1 + 1 = 5 + 1 = 6$. Now, if $\mathbf{d}'$ where graphic, then we would have a graph with five vertices, one of which has a degree of 4 and another that has a degree of 0, and no two vertices have the same degree. Applying either Theorem 2.9 (or its proof), we see that this is not possible. Thus, $\mathbf{d}'$ is not graphic, and so, $\mathbf{d}$ is not graphic.

Remark 2.20. There are several proofs of the next theorem, but they are outside the scope of this text. See Ref. [3] for a short inductive proof.

Theorem 2.21 (Erdös–Gallai theorem). *A degree sequence* $\mathbf{d} = (d_1, \ldots, d_n)$ *is graphic if and only if its sum is even and for all* $1 \leq k \leq n-1$,

$$\sum_{i=1}^{k} d_i \leq k(k+1) + \sum_{i=k+1}^{n-1} \min\{k+1, d_i\}. \tag{2.4}$$

□

2.2 Some Types of Graphs from Degree Sequences

Definition 2.22 (Complete graph). Let $G = (V, E)$ be a graph, with $|V| = n$ with $n \geq 1$. If the degree sequence of G is $(n-1, n-1, \ldots, n-1)$, then G is called a complete graph on n vertices and is denoted by K_n. In a complete graph on n vertices, each vertex is connected to every other vertex by an edge.

Lemma 2.23. *Let* $K_n = (V, E)$ *be the complete graph on* n *vertices. Then,*

$$|E| = \frac{n(n-1)}{2}.$$

□

Corollary 2.24. *Let* $G = (V, E)$ *be a graph, and let* $|V| = n$. *Then,*

$$0 \leq |E| \leq \binom{n}{2}.$$

□

Definition 2.25 (Regular graph). Let $G = (V, E)$ be a graph with $|V| = n$. If the degree sequence of G is $(k, k, \ldots, k)$ with $k \leq n-1$, then G is called a k-regular graph on n vertices.

Example 2.26. We illustrate one complete graph and two (non-complete) regular graphs in Fig. 2.3. Obviously, every complete graph is a regular graph. Every Platonic solid is also a regular graph, but not every regular graph is a Platonic solid. In Fig. 2.3(c), we show a flattened dodecahedron, one of the five platonic solids from classical geometry. The Petersen graph (Fig. 2.3(b)) is a 3-regular graph that is used in many graph-theoretic examples.

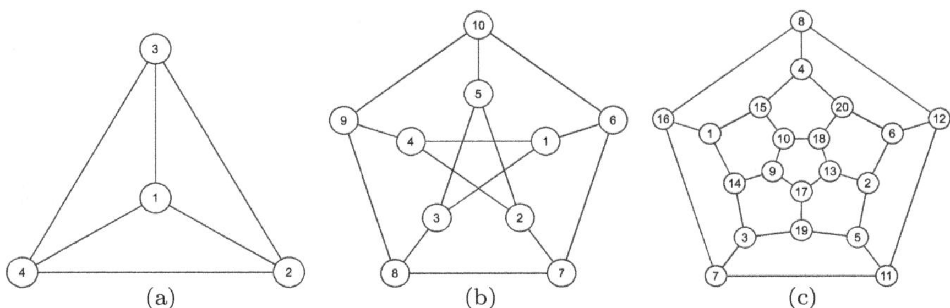

Fig. 2.3 (a) The complete graph K_4, (b) the "Petersen graph," and (c) the dodecahedron. All Platonic solids are three-dimensional representations of regular graphs, but not all regular graphs are Platonic solids. These figures were generated with Maple.

2.3 Subgraphs

Definition 2.27 (Subgraph). Let $G = (V, E)$. A graph $H = (V', E')$ is a *subgraph* of G if $V' \subseteq V$ and $E' \subseteq E$. The subgraph H is *proper* if $V' \subsetneq V$ or $E' \subsetneq E$.

Example 2.28. We illustrate the notion of a subgraph in Fig. 2.4. The Petersen graph is shown on the left. A subgraph containing vertices 1–5 is highlighted in the middle. On the right, we show a subgraph of the Petersen graph containing vertices 6–12 on its own.

Definition 2.29 (Spanning subgraph). Let $G = (V, E)$ be a graph and $H = (V', E')$ be a subgraph of G. The subgraph H is a *spanning subgraph* of G if $V' = V$.

Definition 2.30 (Edge-induced subgraph). Let $G = (V, E)$ be a graph. If $E' \subseteq E$, the subgraph of G induced by E' is the graph $H = (V', E')$, where $v \in V'$ if and only if v appears in an edge in E.

Example 2.31. Using the Petersen graph, we illustrate a subgraph induced by a vertex subset and a spanning subgraph. In Fig. 2.4(b), we illustrate the subgraph induced by the vertex subset $V' = \{1, 2, 3, 4, 5\}$ (highlighted). In Fig. 2.5, we have a spanning

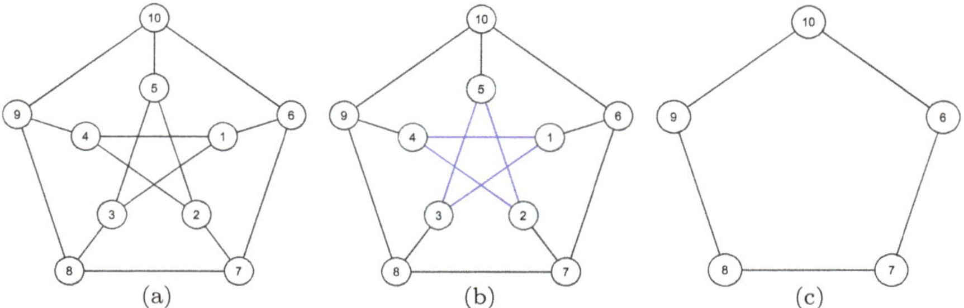

Fig. 2.4 (a) The Petersen graph is shown with (b) a subgraph highlighted and (c) that subgraph displayed on its own. A subgraph of a graph is another graph whose vertices and edges are subcollections of those of the original graph.

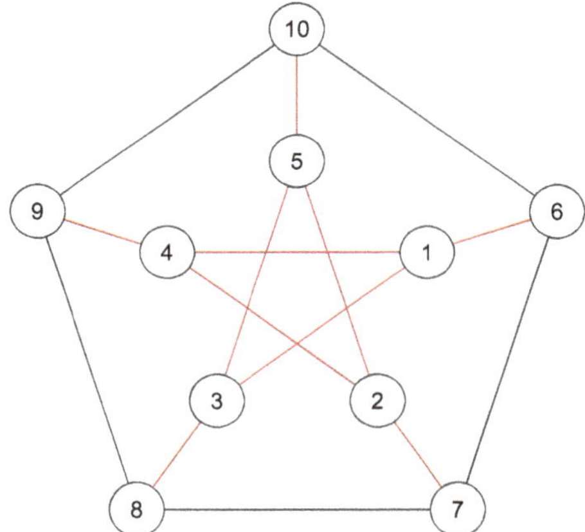

Fig. 2.5 The spanning subgraph is induced by the edge subset $E' = \{\{1,3\}, \{1,4\}, \{1,6\}, \{2,4\}, \{2,5\}, \{2,7\}, \{3,5\}, \{3,8\}, \{4,9\}, \{5,10\}\}$.

subgraph induced by the edge subset

$$E' = \{\{1,3\}, \{1,4\}, \{1,6\}, \{2,4\}, \{2,5\}, \\ \{2,7\}, \{3,5\}, \{3,8\}, \{4,9\}, \{5,10\}\}.$$

Definition 2.32 (Vertex-induced subgraph). Let $G = (V, E)$ be a graph. If $V' \subseteq E$, the subgraph of G induced by V' is the graph $H = (V', E')$, where $\{v_1, v_2\} \in E'$ if and only if v_1 and v_2 are both in V'.

Remark 2.33. For directed graphs, all subgraph definitions are modified in the obvious way. Edges become directed as one would expect.

2.4 Cliques, Independent Sets, Complements, and Covers

Definition 2.34 (Clique). Let $G = (V, E)$ be a graph. A *clique* is a set $S \subseteq V$ of vertices so that:

(1) the subgraph induced by S is a complete graph (or in general graphs, every pair of vertices in S is connected by at least one edge in E); and
(2) if $S' \supset S$, there is at least one pair of vertices in S' that are not connected by an edge in E.

Remark 2.35. There is sometimes a little contention about the definition of clique. Some people define it to be any set of vertices of a graph that induces a complete graph. That is, they drop the second property of Definition 2.34. In this case, a clique S satisfying the second property of Definition 2.34 is called a *maximal clique* because no other vertex can be added to the set while keeping the set a clique. A *maximum cardinality clique* of a graph is simply a clique that is largest in size (number of elements) among all possible cliques of the graph.

Definition 2.36 (Independent set). Let $G = (V, E)$ be a graph. An *independent set* of G is a set $I \subseteq V$ so that no pair of vertices in I is joined by an edge in E. A set $I \subseteq V$ is a *maximal independent* set if I is independent and if there is no other set $J \supset I$ such that J is also independent.

Example 2.37. The easiest way to think of cliques is as subgraphs that are K_n but so that no larger set of vertices induces a larger complete graph. Maximal independent sets are the opposite of cliques. The graph illustrated in Fig. 2.6(a) has three cliques. An independent set is illustrated in Fig. 2.6(b).

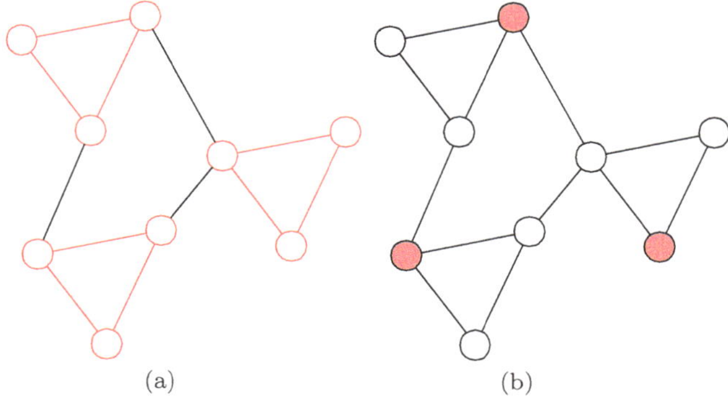

Fig. 2.6 (a) A clique is a set of vertices in a graph that induces a complete graph as a subgraph and so that no larger set of vertices has this property. The graph in this figure has three cliques. (b) An independent set of vertices that are not adjacent is also shown.

Definition 2.38 (Clique number). Let $G = (V, E)$ be a graph. The *clique number* of G, written $\omega(G)$, is the size (number of vertices) of the largest clique in G.

Definition 2.39 (Independence number). The *independence number* of a graph $G = (V, E)$, written $\alpha(G)$, is the size of the largest independent set of G.

Definition 2.40 (Graph complement). Let $G = (V, E)$ be a graph. The *graph complement* of G is a graph $H = (V, E')$ so that

$$e = \{v_1, v_2\} \in E' \qquad \iff \qquad \{v_1, v_2\} \notin E.$$

Example 2.41. In Fig. 2.7, the graph from Fig. 2.6 is illustrated (in a different spatial configuration) with its cliques. The complement of the graph is also illustrated. Note that in the complement, every clique is now an independent set.

Theorem 2.42. *Let $G = (V, E)$ be a graph, and let $H = (V, E')$ be its complement. A set S is a clique in G if and only if S is a maximal independent set in H.* □

Remark 2.43. We may compare the graph complement to the relative complement of a subgraph.

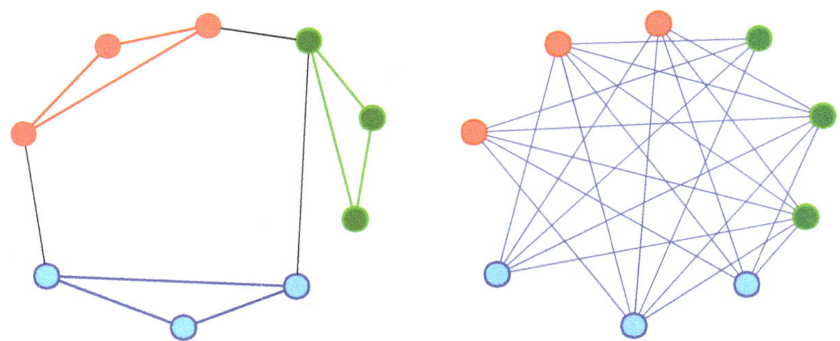

Fig. 2.7 A graph and its complement with cliques in one illustrated and independent sets in the other illustrated.

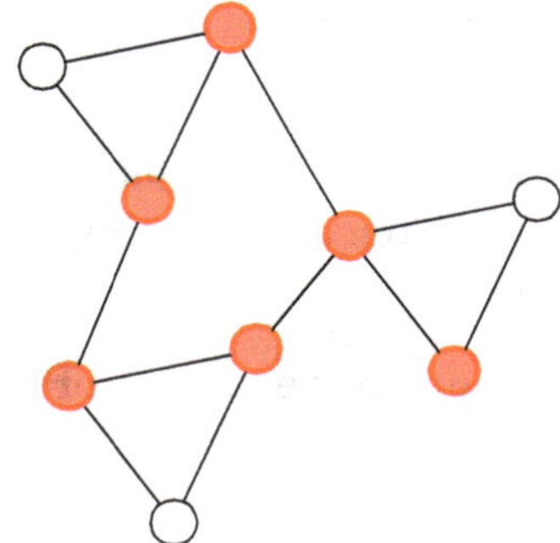

Fig. 2.8 A vertex cover is a set of vertices with the property that every edge has at least one endpoint inside the covering set.

Definition 2.44 (Vertex cover). Let $G = (V, E)$ be a graph. A vertex cover is a set of vertices $S \subseteq V$ so that for all $e \in E$, at least one element of e is in S; i.e., every edge in E is adjacent to at least one vertex in S.

Example 2.45. A vertex cover is illustrated in Fig. 2.8.

Theorem 2.46. *A set I is an independent set in a graph $G = (V, E)$ if and only if the set $V \setminus I$ is a covering in G.*

Proof. ($\Rightarrow$) Suppose that I is an independent set and we choose $e = \{v, v'\} \in E$. If $v \in I$, then clearly $v' \in V \setminus I$. The same is true of v'. It is possible that neither v nor v' is in I, but this does not affect the fact that $V \setminus I$ must be a cover since for every edge $e \in E$, at least one element of e is in $V \setminus I$.

($\Leftarrow$) Now, suppose that $V \setminus I$ is a vertex covering. Choose any two vertices v and v' in I. The fact that $V \setminus I$ is a vertex covering implies that $\{v, v'\}$ cannot be an edge in E because it does not contain at least one element from $V \setminus I$, contradicting our assumption about $V \setminus I$. Thus, I is an independent set since no two vertices in I are connected by an edge in E. This completes the proof. □

Remark 2.47. Theorem 2.46 shows that the problem of identifying a largest independent set is essentially identical to the problem of identifying a minimum (size) vertex covering. To see one example of the utility of a vertex covering, imagine a graph structure defined by the hallways of a building. Vertices are the intersection of two hallways. Finding the *minimal vertex covering* asks the question: What is the minimum number of guards or cameras that must be used to monitor each hallway?

2.5 Chapter Notes

Paul Erdös (1913–1996), also known as the "Magician from Budapest" [4], is the most prolific mathematician in history (after Euler), publishing over 1,400 papers [5]. Erdös was known for his unusual lifestyle and deep love of mathematics. He is an academic brother of John von Neumann, one of the most influential mathematicians of the 20th century, as well as Pál Turán, with whom he collaborated.

Tibor Gallai was also a Hungarian mathematician who worked closely with Paul Erdös. Gallai worked in graph theory and is also known for the Edmonds–Gallai decomposition [6], which is closely related to matching in graphs—which we will study later. His advisor was König, whose results on matchings and coverings we will be studying later.

Václav Havel published a version of the Havel–Hakimi theorem (algorithm) in 1955 [7]. Seifollah Louis Hakimi published his own version in 1962 [8]. Hakimi was an academic descendant of Vannevar Bush, one of the most influential engineers of the 20th century in the United States.

The investigation of graphs with specific degree sequences became popular as a result of the work in network science [9]. *Scale-free* graphs (or networks) have a degree sequence that follows a power-law distribution; that is, the number of vertices with a degree of d, denoted by $n(d)$, behaves according to the law $n(d) \propto d^{-\gamma}$, where γ is called the scaling constant. To find out why all this investigation started, see Barabási and Albert's original work [10], as well as commentaries, e.g., Ref. [11]. A more mathematical (and graph-theoretic) perspective on the properties of graphs with specific degree distributions can be found in Refs. [12–15].

The graph covering problem is closely related to the *dominating set problem*. A *dominating set* in a graph $G = (V, E)$ is a set of vertices $D \subseteq V$ so that every vertex is either in V or adjacent to a vertex in D. An example is the classical *art gallery problem* [16], in which the objective is to position a minimum number of guards to guard an art gallery. The art gallery problem can be reduced to a dominating set problem for a specially designed graph.

2.6 Exercises

Exercise 2.1
Prove Corollary 2.11.

Exercise 2.2
Prove Theorem 2.12.

Exercise 2.3
Prove Corollary 2.15.

Exercise 2.4
Decide whether the following sequence is graphic: $\mathbf{d} = (6, 4, 3, 3, 2, 1, 1)$.

Exercise 2.5
Develop a (recursive) algorithm based on Theorem 2.17 to determine whether a sequence is graphic.

Exercise 2.6
Prove Lemma 2.23 and Corollary 2.24. [Hint: Use Eq. (2.1).]

Exercise 2.7
Assume that the clique number of a graph G is 4. What is the independence number of the complement of G?

Exercise 2.8
Consider the following Petersen graph.

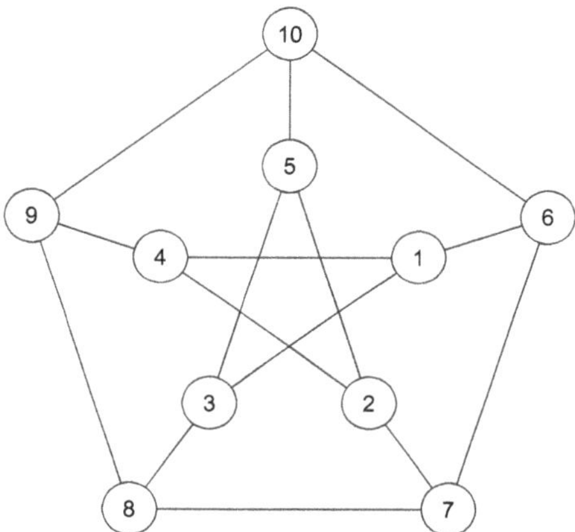

Find a minimal (minimum size) vertex covering for this graph. Can you use this to find a maximal (maximum size) independent set?

Exercise 2.9
Find the clique and independence numbers of the graph shown in Fig. 2.6(a) and (b).

Exercise 2.10
Prove Theorem 2.42. [Hint: Use the definition of graph complement and the fact that if an edge is present in a graph G, it must be absent in its complement.]

Exercise 2.11
Illustrate by exhaustion that removing any vertex from the proposed covering in Fig. 2.8 destroys the covering property.

Chapter 3

Walks, Cycles, Cuts, and Centrality

Remark 3.1 (Chapter goals). In this chapter, we discuss the idea of traversing a graph along its edges in what is called a walk in a graph. We discuss special kinds of walks called Hamiltonian paths and Eulerian trails. Cycles are discussed along with additional graph-theoretic vocabulary. We discuss connected and disconnected graphs and special sets of edges and vertices (called cuts), whose removal disconnects a graph. Applications to centrality (vertex rankings) are also discussed.

3.1 Paths, Walks, and Cycles

Definition 3.2 (Walk). Let $G = (V, E)$ be a graph. A walk $w = (v_1, e_1, v_2, e_2, \ldots, v_n, e_n, v_{n+1})$ in G is an alternating sequence of vertices and edges in V and E, respectively, so that for all $i = 1, \ldots, n$, $\{v_i, v_{i+1}\} = e_i$. A walk is called *closed* if $v_1 = v_{n+1}$ and *open* otherwise. A walk consisting of only one vertex is called *trivial.*

Definition 3.3 (Sub-walk). Let $G = (V, E)$ be a graph. If w is a walk in G, then a *sub-walk* of w is any walk w' that is also a subsequence of w.

Remark 3.4. Let $G = (V, E)$ to each walk

$$w = (v_1, e_1, v_2, e_2, \ldots, v_n, e_n, v_{n+1}).$$

We can associate a subgraph $H = (V', E')$ with:

(1) $V' = \{v_1, \ldots, v_{n+1}\}$; and
(2) $E' = \{e_1, \ldots, e_n\}$.

We call this the subgraph induced by the walk w.

Definition 3.5 (Trail/Tour). Let $G = (V, E)$ be a graph. A trail in G is a walk in which no edge is repeated. A *tour* is a closed trail. A trail or tour is called *Eulerian* if it contains each edge in E exactly once.

Definition 3.6 (Path). Let $G = (V, E)$ be a graph. A *path* in G is a *nontrivial* walk with no vertex and no edge repeated. A *Hamiltonian path* is a path that contains exactly one copy of each vertex in V.

Definition 3.7 (Cycle). A closed walk with a length of at least 3 and no repeated edges and in which the *only* repeated vertices are the first and the last is called a cycle. A cycle in a graph is *Hamiltonian* if it contains every vertex.

Definition 3.8 (Hamiltonian and Eulerian graphs). A graph $G = (V, E)$ is said to be *Hamiltonian* if it contains a Hamiltonian cycle and Eulerian if it contains an Eulerian tour.

Example 3.9. We illustrate an Eulerian trail and a Hamiltonian path in Fig. 3.1. Note that Vertex 1 is repeated in the trail, meaning that this is not a *path.* We contrast this with Fig. 3.3(b), which shows a Hamiltonian path. Here, each vertex occurs exactly once in the illustrated path, but not all the edges are included. In this graph, it is impossible to have either a Hamiltonian cycle or an Eulerian tour.

Example 3.10. An Eulerian graph and a separate Hamiltonian graph are shown in Fig. 3.2. The edges are numbered in the order one could use to show the tour or the cycle. Note that vertices are reused in the tour, while only the first and last vertices are the same in the Hamiltonian cycle (as required).

Remark 3.11. We will return to the discussion of Eulerian graphs in Section 4.3.

Definition 3.12 (Length). The *length* of a walk is the number of edges contained in it.

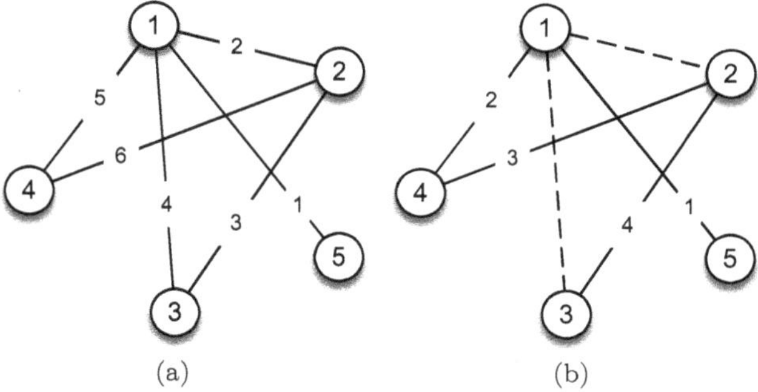

Fig. 3.1 (a) An Eulerian trail and (b) a Hamiltonian path are illustrated.

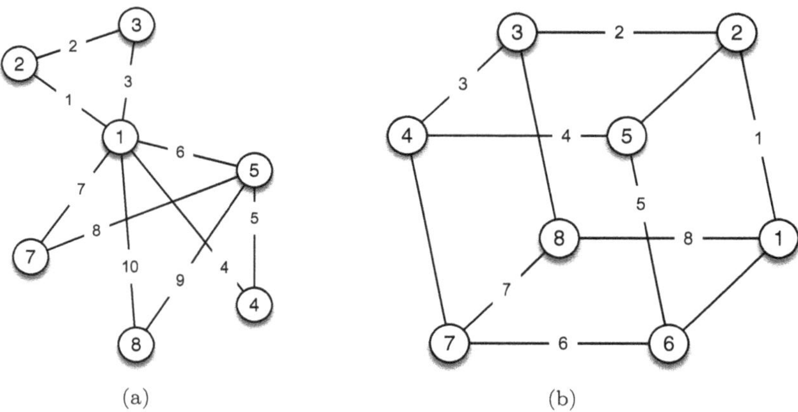

Fig. 3.2 (a) An Eulerian tour and (b) a Hamiltonian cycle are illustrated. These graphs are Eulerian and Hamiltonian, respectively.

Example 3.13. A walk is illustrated in Fig. 3.3(a). Formally, this walk can be written as

$$w = (1, \{1, 4\}, 4, \{4, 2\}, 2, \{2, 3\}, 3).$$

The cycle shown in Fig. 3.3(b) can be formally written as

$$c = (1, \{1, 4\}, 4, \{4, 2\}, 2, \{2, 3\}, 3, \{3, 1\}, 1).$$

Note that the cycle begins and ends with the same vertex (that's what makes it a cycle). Also, w is a sub-walk of c. Note further that

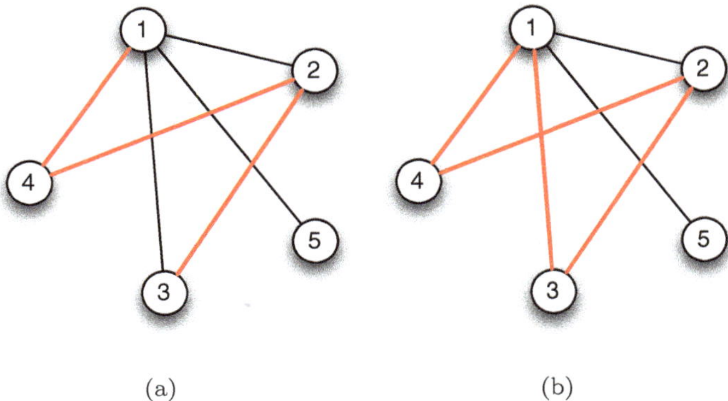

(a) (b)

Fig. 3.3 (a) A walk and (b) a cycle are illustrated.

we could easily have represented the walk as

$$w = (3, \{3, 2\}, 2, \{2, 4\}, 4, \{4, 1\}, 1).$$

In general, we can shift the ordering of the cycle in anyway (for example, beginning at Vertex 2). Thus, we see that in an undirected graph, a cycle or walk representation may not be unique.

Remark 3.14. If w is a path in a graph $G = (V, E)$, then the subgraph induced by w is simply the graph composed of the vertices and edges in w.

Definition 3.15 (Path graph). Suppose that $G = (V, E)$ is a graph with $|V| = n$. If w is a Hamiltonian path in G and H is the subgraph induced by w and $H = G$, then G is called a *n-path* or a *path graph* on n vertices, denoted by P_n.

Definition 3.16 (Cycle graph). If w is a Hamiltonian cycle in G and H is the subgraph induced by w and $H = G$, then G is called a *n-cycle* or a *cycle graph* on n vertices, denoted by C_n.

Example 3.17. We illustrate a cycle graph with six vertices (6-cycle or C_6) and a path graph with four vertices (4-path or P_4) in Fig. 3.4.

Remark 3.18. Walks, cycles, paths, and tours can all be extended to the case of digraphs. In this case, the walk, path, cycle, or tour *must* respect the edge directionality. Thus, if $w = (\ldots, v_i, e_i, v_{i+1}, \ldots)$ is a

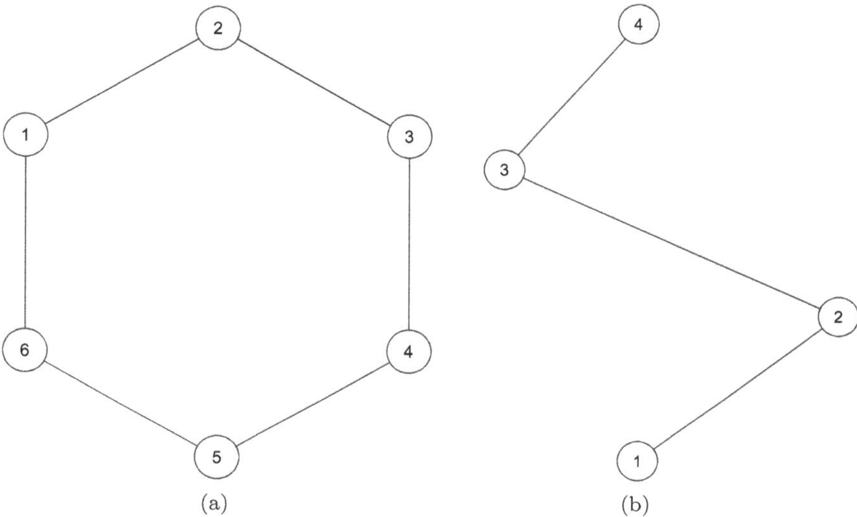

Fig. 3.4 (a) 6-cycle and (b) 4-path.

directed walk, then $e_i = (v_i, v_{i+1})$ in an ordered pair in the edge set of the graph.

3.2 More Graph Properties: Diameter, Radius, Circumference, and Girth

Definition 3.19 (Distance). Let $G = (V, E)$. The *distance* between v_1 and v_2 in V is the length of the *shortest* walk beginning at v_1 and ending at v_2 if such a walk exists. Otherwise, it is $+\infty$. We write $d_G(v_1, v_2)$ for the distance from v_1 to v_2 in G.

Definition 3.20 (Directed distance). Let $G = (V, E)$ be a digraph. The *(directed) distance* between v_1 to v_2 in V is the length of the *shortest* directed walk beginning at v_1 and ending at v_2 if such a walk exists. Otherwise, it is $+\infty$.

Definition 3.21 (Diameter). Let $G = (V, E)$ be a graph. The *diameter* of G, denoted by $\text{diam}(G)$, is the length of the largest distance in G. That is,

$$\mathrm{diam}(G) = \max_{v_1, v_2 \in V} d_G(v_1, v_2). \tag{3.1}$$

Definition 3.22 (Eccentricity). Let $G = (V, E)$, and let $v_1 \in V$. The *eccentricity* of v_1 is the largest distance from v_1 to any other vertex v_2 in V. That is,

$$\mathrm{ecc}(v_1) = \max_{v_2 \in V} d_G(v_1, v_2). \tag{3.2}$$

Remark 3.23. Naturally, we can define diameter in terms of eccentricity as

$$\mathrm{diam}(G) = \max_{v \in V} \mathrm{ecc}(v). \tag{3.3}$$

Definition 3.24 (Radius). Let $G = (V, E)$. The *radius* of G is the minimum eccentricity of any vertex in V. That is,

$$\mathrm{rad}(G) = \min_{v_1 \in V} \mathrm{ecc}(v_1) = \min_{v_1 \in V} \max_{v_2 \in V} d_G(v_1, v_2). \tag{3.4}$$

Definition 3.25 (Girth). Let $G = (V, E)$ be a graph. If there is a cycle in G (that is, G has a cycle-graph as a subgraph), then the *girth* of G is the length of the *shortest* cycle. When G contains no cycle, the girth is defined as 0.

Definition 3.26 (Circumference). Let $G = (V, E)$ be a graph. If there is a cycle in G (that is, G has a cycle-graph as a subgraph), then the *circumference* of G is the length of the *longest* cycle. When G contains no cycle, the circumference is defined as $+\infty$.

Example 3.27. The eccentricities of the vertices of the graph shown in Fig. 3.5 are:

(1) Vertex 1: 1,
(2) Vertex 2: 2,
(3) Vertex 3: 2,
(4) Vertex 4: 2, and
(5) Vertex 5: 2.

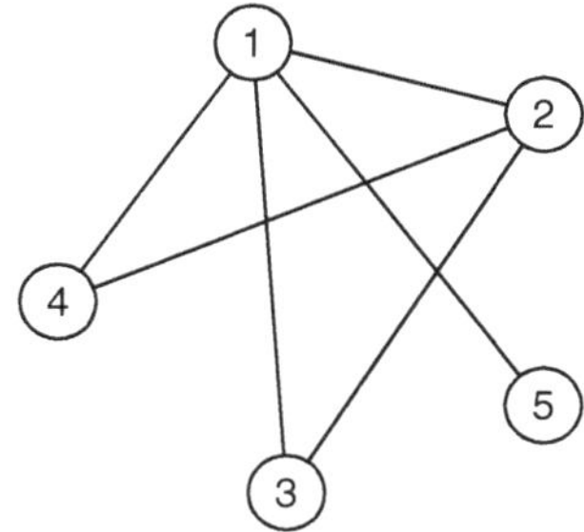

Fig. 3.5 A graph with a diameter of 2, radius of 1, girth of 3, and circumference of 4.

This means that the diameter of the graph is 2 and its radius is 1. We have already seen that there is a 4-cycle subgraph in the graph — see Fig. 3.3(b). This is the largest cycle in the graph, so the circumference of the graph is 4. There are several 3-cycles in the graph (an example being the cycle $(1, \{1, 2\}, 2, \{2, 4\}, 4, \{4, 1\}, 1)$). The smallest possible cycle is a 3-cycle. Thus, the girth of the graph is 3.

3.3 Graph Components

Definition 3.28 (Reachability). Let $G = (V, E)$, and let v_1 and v_2 be two vertices in V. Then, v_2 is *reachable* from v_1 if there is a walk w beginning at v_1 and ending at v_2 (alternatively, the distance from v_1 to v_2 is not $+\infty$). If G is a digraph, we assume that the walk is directed.

Definition 3.29 (Connectedness). A graph G is connected if for every pair of vertices v_1 and v_2 in V, v_2 is reachable from v_1. If G is a digraph, then G is connected if its *underlying graph* is connected. A graph that is not connected is called *disconnected.*

Definition 3.30 (Strong connectedness). A digraph G is strongly connected if for every pair of vertices v_1 and v_2 in V, v_2 is reachable (by a directed walk) from v_1.

Remark 3.31. In Definition 3.30, we are really requiring, for any pair of vertices v_1 and v_2, that v_1 be reachable from v_2 *and* v_2 be

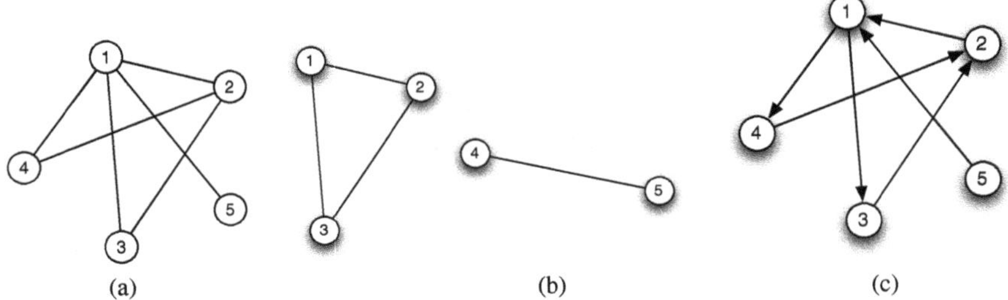

Fig. 3.6 (a) A connected graph, (b) a disconnected graph with two components, and (c) a connected digraph that is not strongly connected.

reachable from v_1 by directed walks. If this is not possible, then a directed graph could be connected but not *strongly connected.*

Example 3.32. In Fig. 3.6, we illustrate a connected graph, a disconnected graph, and a connected digraph that is not strongly connected.

Definition 3.33 (Component). Let $G = (V, E)$ be a graph. A subgraph H of G is a *component* of G if:

(1) H is connected; and
(2) K is a subgraph of G and H is a proper subgraph of K, then K is not connected. The number of components of a graph G is written as $c(G)$.

Remark 3.34. A component H of a graph G is a *maximally connected subgraph of G*. Here, *maximal* is taken with respect to the subgraph ordering. That is, H is less than K in the subgraph ordering if H is a proper subgraph of K.

Example 3.35. Figure 3.6(b) contains two components: 3-cycle and 2-path.

Proposition 3.36. *A connected graph G has exactly one component.*

□

Definition 3.37 (Edge deletion graph). Let $G = (V, E)$, and let $E' \subseteq E$. Then, the graph G' resulting from deleting the edges in E' from G is the subgraph induced by the edge set $E \setminus E'$. We write this as $G' = G - E'$.

Definition 3.38 (Vertex deletion graph). Let $G = (V, E)$, and let $V' \subseteq V$. Then, the graph G' resulting from deleting the edges in V' from G is the subgraph induced by the vertex set $V \setminus V'$. We write this as $G' = G - V'$.

Definition 3.39 (Vertex cut and cut vertex). Let $G = (V, E)$ be a graph. A set $V' \subset V$ is a *vertex cut* if the graph G' resulting from deleting vertices V' from G has *more components* than graph G. If $V' = \{v\}$ is a vertex cut, then v is called a *cut vertex.*

Definition 3.40 (Edge cut and cut edge). Let $G = (V, E)$ be a graph. A set $E' \subset E$ is an *edge cut* if the graph G' resulting from deleting edges E' from G has *more components* than graph G. If $E' = \{e\}$ is an edge cut, then e is called a *cut edge.*

Definition 3.41 (Minimal edge cut). Let $G = (V, E)$. An edge cut E' of G is *minimal* if when we remove any edge from E' to form E'', the new set E'' is no longer an edge cut.

Example 3.42. In Fig. 3.7, we illustrate a vertex cut and a cut vertex (a singleton vertex cut) and an edge cut and a cut edge (a singleton edge cut). Note that the edge cut in Fig. 3.7(a) is minimal and cut edges are always minimal. A cut edge is sometimes called a *bridge* because it connects two distinct components in a graph. Bridges (and small edge cuts) are a very important part of social network analysis [17, 18] because they represent connections

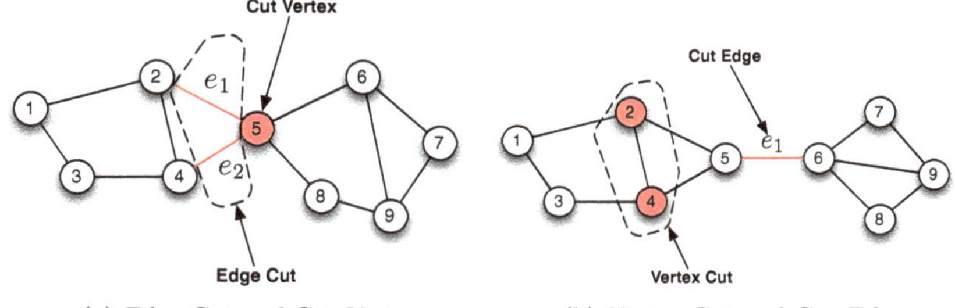

(a) Edge Cut and Cut Vertex (b) Vertex Cut and Cut Edge

Fig. 3.7 Illustrations of a vertex cut and a cut vertex (a singleton vertex cut) and an edge cut and a cut edge (a singleton edge cut). Cuts are sets of vertices or edges whose removal from a graph creates a new graph with more components than the original graph.

between different communities. To see this, suppose that Fig. 3.7(b) represents the communication patterns between two groups. The fact that Members 5 and 6 communicate *and* that these are the only two individuals who communicate between these two groups could be important in understanding how these groups interrelate.

Theorem 3.43. *Let $G = (V, E)$ be a connected graph, and let $e \in E$. Then, $G' = G - \{e\}$ is connected if and only if e lies on a cycle in G.*

Proof. ($\Leftarrow$) Recall that a graph G is connected if and only if for every pair of vertices v_1 and v_{n+1}, there is a walk w from v_1 to v_{n+1} with

$$w = (v_1, e_1, v_2, \ldots, v_n, e_n, v_{n+1}).$$

Let $G' = G - \{e\}$. Suppose that e lies on a cycle c in G, and choose two vertices v_1 and v_{n+1} in G. If e is not on any walk w connecting v_1 to v_{n+1} in G, then the removal of e does not affect the reachability of v_1 and v_{n+1} in G'. Therefore, assume that e is in the walk w. The fact that e is in a cycle of G implies that we have vertices $u_1, \ldots, u_m$ and edges $f_1, \ldots, f_m$ so that

$$c = (u_1, f_1, \ldots, u_m, f_m, u_1)$$

is a cycle and e is among $f_1, \ldots, f_m$. Without loss of generality, assume that $e = f_m$ and that $e = \{u_m, u_1\}$. (Otherwise, we can reorder the cycle to make this true.) Then, in G', we have the path

$$c' = (u_1, f_1, \ldots, f_{m-1}, u_m).$$

The fact that e is in the walk w implies that there are vertices v_i and v_{i+1} so that $e = \{v_i, v_{i+1}\}$ (with $v_i = u_1$ and $v_{i+1} = u_m$). In deleting e from G, we remove the sub-walk (v_i, e, v_{i+1}) from w. However, we can create a new walk with the structure

$$w' = (v_1, e_1, \ldots, v_i, f_1, u_2, \ldots, u_{m-1}, f_{m-1}, u_m, \ldots, e_n, v_{n+1}).$$

This is illustrated in Fig. 3.8.

($\Rightarrow$) Suppose $G' = G - \{e\}$ is connected. Now, let $e = \{v_1, v_{n+1}\}$. Since G' is connected, there is a walk from v_1 to v_{n+1}. Applying

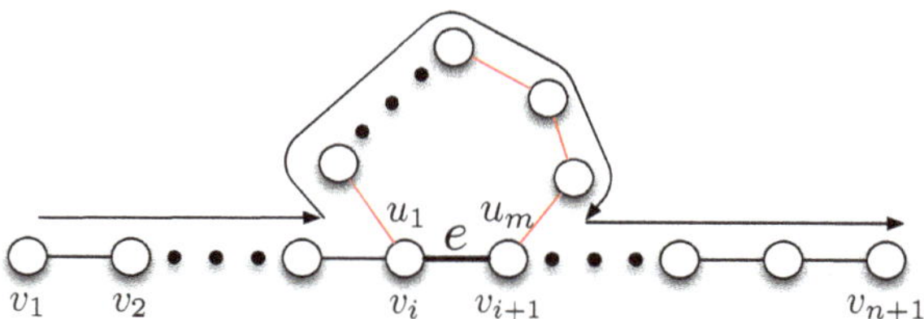

Fig. 3.8 If e lies on a cycle, then we can repair path w by *going the long way around the cycle* to reach v_{n+1} from v_1.

Remark 4.28, we can reduce this walk to a path p, with

$$p = (v_1, e_1, \ldots, v_n, e_n, v_{n+1}).$$

Since p is a path, there are no repeated vertices in p. We can construct a cycle c containing e in G as

$$p = (v_1, e_1, \ldots, v_n, e_n, v_{n+1}, e, v_1)$$

since $e = \{v_1, v_{n+1}\} = \{v_{n+1}, v_1\}$. Thus, e lies on a cycle in G. This completes the proof. □

Corollary 3.44. *Let $G = (V, E)$ be a connected graph, and let $e \in E$. The edge e is a cut edge if and only if e does not lie on a cycle in G.* □

Definition 3.45 (k-connected). A graph is k connected if it has at least k vertices and if the graph remains connected after removing fewer than k vertices (to form the vertex deletion graph).

Remark 3.46. We pick up the study of connectivity when we discuss flows and cuts on graphs in Chapter 6. In particular, k-connectivity is an important part of building reliable networks [19].

Remark 3.47. The following result is taken from extremal graph theory, the study of extremes or bounds in the properties of graphs. There are a number of results in extremal graph theory that are of interest. See Ref. [20] for a complete introduction.

Theorem 3.48. *If $G = (V, E)$ is a graph with n vertices and k components, then*

$$|E| \leq \frac{(n - k + 1)(n - k)}{2}. \tag{3.5}$$

Proof. Assume that each component of G has n_i vertices in it, with $\sum_{i=1}^{k} n_i = n$. Applying Lemma 2.23, we know that component i has at most $n_i(n_i - 1)/2$ edges; that is, each component is a complete graph on n_i vertices. This is the largest number of edges that can occur under these assumptions.

Consider the case where $k - 1$ of the components have exactly one vertex and the remaining components have $n - (k - 1)$ vertices. Then, the total number of edges in this case is at most

$$\frac{(n-(k-1))(n-(k-1)-1)}{2} = \frac{(n-k+1)(n-k)}{2}.$$

It now suffices to show that this case has the highest number of vertices of all cases where the k components are each complete graphs.

Consider the case when component i is K_r and component j is K_s, with $r, s \geq 2$, and suppose $r \geq s$. Then, the total number of edges in *these two components* is

$$\frac{r(r-1)+s(s-1)}{2} = \frac{r^2+s^2-r-s}{2}.$$

Now, suppose we move one vertex in component j to component i. Then, component i is now K_{r+1} and component j is now K_{s-1}. Applying Lemma 2.23, the number of edges in this case is

$$\frac{(r+1)(r)+(s-1)(s-2)}{2} = \frac{r^2+r+s^2-3s+2}{2}.$$

Observe that since $r \geq s$, substituting s for r, we have

$$r^2+r+s^2-3s+2 \geq r^2+s^2-2s+2.$$

By a similar argument,

$$r^2+s^2-2s \geq r^2+s^2-r-s.$$

Thus, we conclude that

$$\begin{aligned}\frac{r^2+r+s^2-3s+2}{2} &\geq \frac{r^2+s^2-2s+2}{2} \geq \frac{r^2+s^2-2s}{2}\\ &\geq \frac{r^2+s^2-r-s}{2}.\end{aligned}$$

Repeating this argument over and over shows that in a k-component graph with n vertices, the largest number of edges must occur in the case when there is one complete component with $n-(k-1)$ vertices and $k-1$ components with exactly one vertex. Thus, the number of edges in a graph with k components must satisfy Eq. (3.5). This completes the proof. □

Corollary 3.49. *Any graph with n vertices and more than* $(n-1)(n-2)/2$ *edges is connected.* □

3.4 Introduction to Centrality

Remark 3.50. There are many situations in which we'd like to measure the importance of a vertex in a graph. The problem of measuring this quantity is usually called determining a vertex's *centrality.*

Definition 3.51 (Degree centrality). Let $G = (V, E)$ be a graph. The *degree centrality* of a vertex is just its degree. That is,

$$C_D(v) = \deg(v).$$

These values can be normalized to lie in the interval $[0, 1]$ by using

$$\hat{C}_D(v) = \frac{\deg(v)}{2|E|}.$$

Remark 3.52. Degree centrality is only the simplest measurement of centrality. There are many other measures of this quantity. We discuss one more and then continue our discussion of this topic in Chapter 10.

Definition 3.53 (Geodesic centrality). Let $G = (V, E)$ be a graph. The *geodesic centrality* (sometimes called the *betweenness centrality*) of a vertex $v \in V$ is the fraction of times v occurs on any shortest path connecting any other pair of vertices $s, t \in V$. Put more formally, let σ_{st} be the total number of shortest paths connecting vertex s to vertex t. Let $\sigma_{st}(v)$ be the number of these shortest paths containing v. The geodesic centrality of v is

$$C_B(v) = \sum_{s \neq t \neq v} \frac{\sigma_{st}(v)}{\sigma_{st}}. \tag{3.6}$$

These values can be normalized so that they fall within $[0, 1]$ by dividing each $C_B(v)$ by the sum of all $C_B(v)$ so that

$$\hat{C}_B(v) = \frac{C_B(v)}{\sum_{v \in V} C_B(v)}.$$

Example 3.54. Consider the graph with four vertices shown in Fig. 3.9. The degrees of the graph are $(2, 3, 3, 2)$, which is the unnormalized degree centrality. The normalized degree centrality of the vertices is

$$\hat{C}_D(v_1) = \frac{1}{5} \qquad \hat{C}_D(v_2) = \frac{3}{10}$$
$$\hat{C}_D(v_3) = \frac{3}{10} \qquad \hat{C}_D(v_4) = \frac{1}{5}.$$

To compute the normalized geodesic centrality, we must compute the fraction of times a vertex appears in a shortest path. This is shown in Table 3.1. In the vertex pair $(1, 2)$, there is exactly one shortest path connecting 1 to 2. Since 1 and 2 are the end points, they are not counted. Vertices 3 and 4 do not appear in this shortest path, so they each receive a zero. For $(1, 4)$, there are **two** shortest paths (one through 2 and the other through 3); therefore, half of the shortest paths contain vertex 2 and half of the shortest paths contain vertex 3. The remainder of the table is filled out in exactly

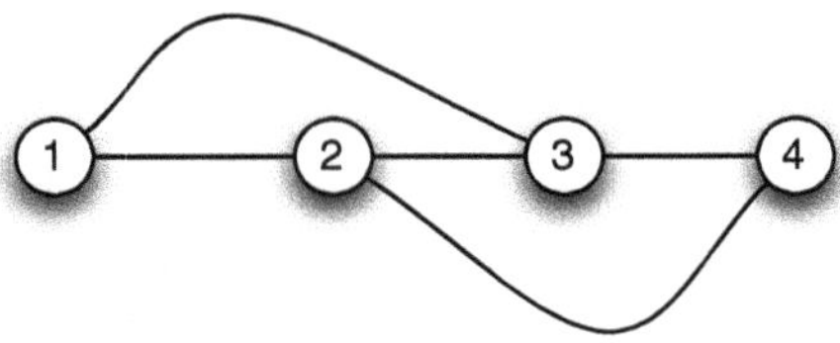

Fig. 3.9 Graph with four vertices.

Table 3.1 A table showing the intermediate computations for geodesic centrality.

Vertex Pair	1	2	3	4
(1,2)	—	—	0	0
(1,3)	—	0	—	0
(1,4)	—	$\frac{1}{2}$	$\frac{1}{2}$	—
(2,3)	0	—	—	0
(2,4)	0	—	0	—
(3,4)	0	0	—	—
SUM	0	$\frac{1}{2}$	$\frac{1}{2}$	0

the same way. The normalized geodesic centrality is

$$\hat{C}_B(v_1) = 0 \quad \hat{C}_B(v_2) = \frac{1}{2}$$

$$\hat{C}_B(v_3) = \frac{1}{2} \quad \hat{C}_B(v_4) = 0.$$

In this case, we see that the geodesic centrality is similar in its ordering of the degree centrality but different in its values.

Remark 3.55. It's clear from this analysis that cut vertices should have high geodesic centrality if they connect two large components of a graph. Thus, by some measures, cut vertices are very important elements of graphs.

3.5 Chapter Notes

Hamiltonian cycles (and paths) are named after William Rowan Hamilton, the most famous Irish mathematician. Most math students know Hamilton for his work on the quaternions [1]. Most physics students know Hamilton for his invention of Hamiltonian mechanics [21], which rephrased Newton's laws in a much more general language, helping to pave the way for quantum mechanics. Hamiltonian cycles are named after Hamilton as a result of his invention of the *icosian game*, which involves finding Hamiltonian cycles on the edges of a dodecahedron. In reality, Hamiltonian paths (as they came to be

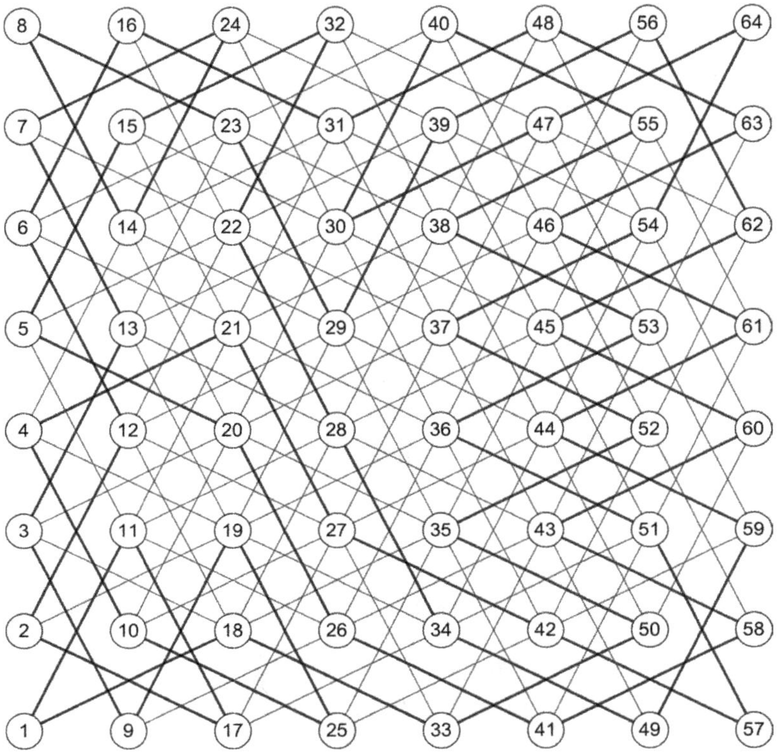

Fig. 3.10 A knight's tour showing that a knight can transit every chess square making only legal moves and returning to its starting position.

known) were studied decades earlier by both Kirkman [22] and (of course) Euler in the form of knight's tours on chessboards [23]. In a knight's tour problem, the graph is composed of vertices given by chess squares, and the set of legal moves a knight may make forms the edge set. This is illustrated in Fig. 3.10

Extremal graph theory is a subfield within graph theory dealing with the relationship between local and global structures in graphs. The field was first investigated by Mantel and Turán [20]. Naturally, Erdös made a contribution as well [24].

Work in social networks is largely distinct from graph theory as a mathematical study. Zachary's karate club [25] is a common first example of the use of graph theory in the study of social networks. However, the use of centrality in studying social networks predates this by at least two decades with the work of Katz [26], who defined "Katz centrality," which is similar to PageRank centrality, which we will discuss in later chapters. PageRank was (of course) created by Brin and Page [27] and formed the basis for the Google search engine.

3.6 Exercises

Exercise 3.1
Prove that it is not possible for a Hamiltonian cycle or Eulerian tour to exist in the graph in Fig. 3.3(a); i.e., prove that the graph is neither Hamiltonian nor Eulerian.

Exercise 3.2
Suppose a graph is Hamiltonian. Does it have a cut vertex? Explain your answer.

Exercise 3.3
Formally define for yourself: directed walks, directed cycles, directed paths, and directed tours for directed graphs. [Hint: Begin with Definition 3.2 and make appropriate changes. Then, do this for cycles, tours, etc.]

Exercise 3.4
Show that Eq. (3.3) is true.

Exercise 3.5
Compute the diameter, radius, girth, and circumference of the Petersen graph.

Exercise 3.6
In an online social network, Alice is friends with Bob and Charlie. Charlie is friends with David and Edward. Edward is friends with Bob:

(1) Find a maximal (maximum size) independent set in the resulting social network graph. What can you interpret from such a set on a social network?
(2) Find the diameter of the social network graph.

Exercise 3.7
Prove Proposition 3.36.

Exercise 3.8
What are the sizes of the smallest vertex and edge cuts in the cycle graph C_k?

Exercise 3.9
Prove Corollary 3.44.

Exercise 3.10
Prove Corollary 3.49.

Exercise 3.11
Compute the geodesic centrality and the degree centrality for the graph shown in Fig. 3.11. Compare your results.

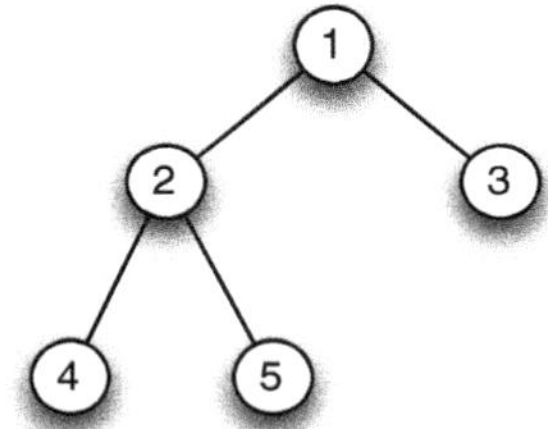

Fig. 3.11 The graph for which centralities are to be computed.

Chapter 4

Bipartite, Acyclic, and Eulerian Graphs

Remark 4.1 (Chapter goals). In this chapter, we build on our work on walks and paths to discuss special graphs: bipartite graphs, acyclic graphs, and Eulerian graphs. We prove results that characterize these graphs. Special attention will be paid to trees, which are critical for computer science and game theory.

4.1 Bipartite Graphs

Definition 4.2. A graph $G = (V, E)$ is *bipartite* if $V = V_1 \cup V_2$ and $V_1 \cap V_2 = \emptyset$, and if $e \in E$, then $e = \{v_1, v_2\}$, with $v_1 \in V_1$ and $v_2 \in V_2$. This definition is valid for non-simple graphs as well.

Remark 4.3. In a bipartite graph, we can think of the vertices as belonging to one of two classes (either V_1 or V_2) and edges only existing between elements of the two classes, not between elements in the same class. We can also define n-partite graphs, in which the vertices are in any of n classes and there are only edges between classes, not within classes.

Example 4.4. Figure 4.1 shows a bipartite graph in which $V_1 = \{1, 2, 3\}$ and $V_2 = \{4, 5\}$. Note that there are only edges connecting vertices in V_1 and vertices in V_2. There are no edges connecting elements in V_1 to other elements in V_1 or elements in V_2 to other elements in V_2.

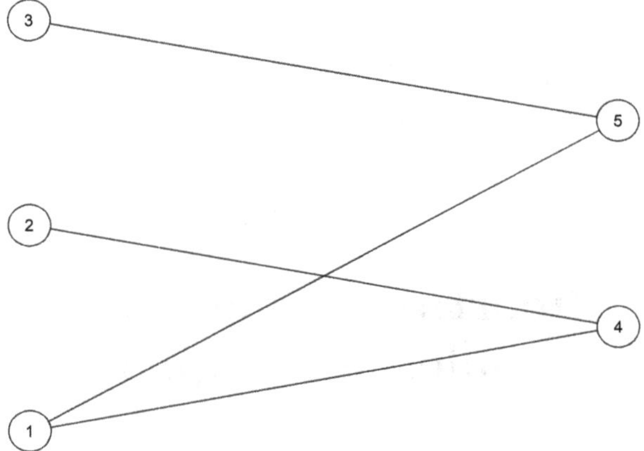

Fig. 4.1 A bipartite graph has two classes of vertices, and edges in the graph only exist between elements of different classes.

Definition 4.5 (Complete bipartite graph). The graph $K_{m,n}$ is the complete bipartite graph consisting of the vertex set $V = \{v_{11}, \ldots, v_{1m}\} \cup \{v_{21}, \ldots, v_{2n}\}$ and having an edge connecting every element of V_1 to every element of V_2.

Definition 4.6 (Path concatenation). Let $p_1 = (v_1, e_1, v_2, \ldots, v_n, e_n, v_{n+1})$ and $p_2 = (v_{n+1}, e_{n+1}, v_{n+2}, \ldots, v_{n+m}, e_{n+m}, v_{n+m+1})$. Then, the *concatenation* of path p_1 with path p_2 is the path

$$p = (v_1, e_1, v_2, \ldots, v_n, e_n, v_{n+1}, e_{n+1}, v_{n+2}, \ldots, v_{n+m}, e_{n+m}, v_{n+m+1}).$$

Remark 4.7. Path concatenation is used in the proof of Theorem 4.8.

Theorem 4.8. *A graph $G = (V, E)$ is bipartite if and only if every cycle in G has an even length.*

Proof. ($\Rightarrow$) Suppose G is bipartite. Every cycle begins and ends at the same vertex and, therefore, in either V_1 or V_2. Without loss of generality, suppose we start with V_1. Starting at a vertex $v_1 \in V_1$, we must take a walk of length 2 to return to V_1. The same is true if we start at a vertex in V_2. Thus, every cycle must contain an even number of edges in order to return to either V_1 or V_2.

($\Leftarrow$) Suppose that every cycle in G has an even length. Without loss of generality, assume G is connected. We create a partition of V so that $V = V_1 \cup V_2$ and $V_1 \cap V_2 = \emptyset$, and there is no edge between vertices if they are in the same class.

Choose an arbitrary vertex $v \in V$, and define

$$V_1 = \{v' \in V : d_G(v, v') \equiv 0 \mod 2\} \quad \text{and} \tag{4.1}$$

$$V_2 = \{v' \in V : d_G(v, v') \equiv 1 \mod 2\}. \tag{4.2}$$

Clearly, V_1 and V_2 constitute a partition of V. Choose $u_1, u_2 \in V_1$, and suppose there is an edge $e = \{u_1, u_2\} \in E$. The distance from v to u_1 is even, so there is a path p_1 with an even number of edges beginning at v and ending at u_1. Likewise, the distance from v to u_2 is even, so there is a path p_2 beginning at u_2 and ending at v with an even number of edges. If we concatenate paths p_1 and the length 1 path $q = (u_1, \{u_1, u_2\}, u_2)$ and path p_2, we obtain a cycle in G that has an odd length (see Fig. 4.2). Therefore, there can be no edge connecting two vertices in V_1.

Choose $u_1, u_2 \in V_2$, and suppose there is an edge $e = \{u_1, u_2\} \in E$. Using the same argument, there is a path p_1 of odd length from v to u_1 and a path p_2 of odd length from u_2 to v. If we concatenate paths p_1 and the length 1 path $q = (u_1, \{u_1, u_2\}, u_2)$ and path p_2, we again obtain a cycle in G that has an odd length. Therefore, there can be no edge connecting two vertices in V_2.

In the case when G has more than one component, execute the process described above for each component to obtain partitions

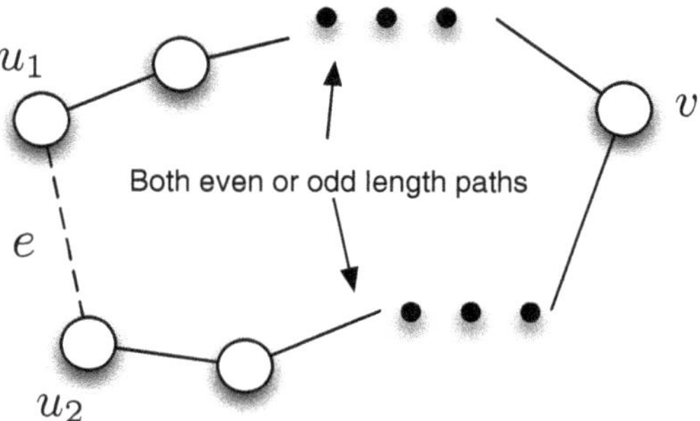

Fig. 4.2 Illustration of the main argument in the proof that a graph is bipartite if and only if all cycles have even length.

$V_1, V_2, V_3, V_4, \ldots, V_{2n}$. Create a bipartition U_1 and U_2 of V with

$$U_1 = \bigcup_{k=1}^{n} V_{2k-1} \quad \text{and} \tag{4.3}$$

$$U_2 = \bigcup_{k=1}^{n} V_{2k}. \tag{4.4}$$

There can be no edge connecting a vertex in U_1 with a vertex in U_2. Thus, G is bipartite. This completes the proof. □

4.2 Acyclic Graphs and Trees

Definition 4.9 (Acyclic graph). A graph that contains *no* cycles is called *acyclic.*

Definition 4.10 (Forests and trees). Let $G = (V, E)$ be an acyclic graph. If G has more than one component, then G is called a *forest.* If G has one component, then G is called a *tree.*

Example 4.11. A randomly generated tree with 10 vertices is shown in Fig. 4.3. Note that the tree (if drawn upside down) can be made to look like a real tree growing up from the ground.

Proposition 4.12. *Every tree is a bipartite graph.* □

Remark 4.13. We can define directed trees and directed forests as acyclic directed graphs. Generally, we require the underlying graphs to be acyclic rather than just having no directed cycles. Directed trees are used frequently in computer science [28], operations research [29], and game theory [30, 31]. For the remainder of this chapter, we deal with undirected trees, but the results presented will also apply to directed trees unless otherwise stated.

Definition 4.14 (Spanning forest). Let $G = (V, E)$ be a graph. If $F = (V', E')$ is an acyclic subgraph of G such that $V = V'$, then F is called a *spanning forest* of G. If F has exactly one component, then F is called a *spanning tree.*

Example 4.15. A spanning tree for the Petersen graph is illustrated in Fig. 4.4. Since the Petersen graph is connected, it is easy to see that

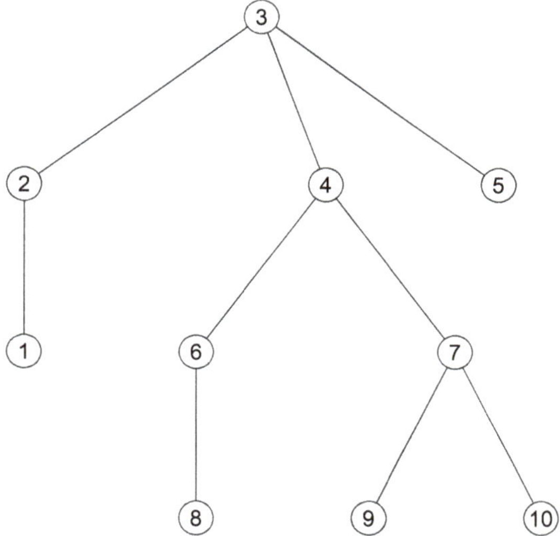

Fig. 4.3 A tree is shown. Imagining the tree upside down illustrates the tree-like nature of the graph structure.

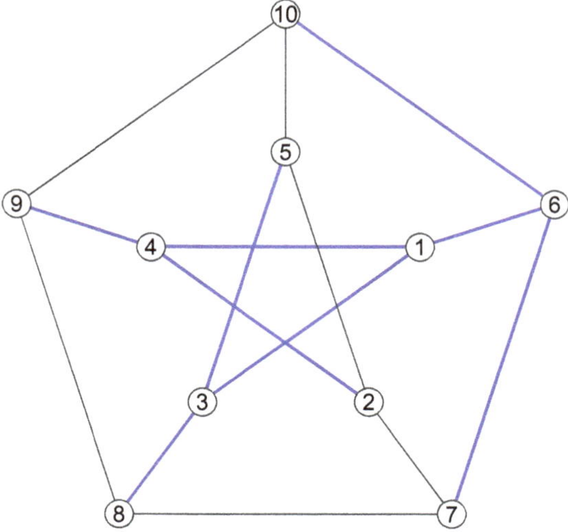

Fig. 4.4 The Petersen graph with a spanning tree shown below.

we do not *need* a spanning forest to construct an acyclic spanning subgraph.

Theorem 4.16. *If $G = (V, E)$ is a connected graph, then there is a spanning tree $T = (V, E')$ of G.*

Proof. We proceed by induction on the number of vertices in G. If $|V| = 1$, then G is itself a (degenerate) tree and thus has a spanning tree. Now, suppose that the statement is true for all graphs G with $|V| \leq n$. Consider a graph G with $n + 1$ vertices. Choose an arbitrary vertex v_{n+1} and remove it and all edges of the form $\{v, v_{n+1}\}$ from G to form G' with vertex set $V' = \{v_1, \ldots, v_n\}$. The graph G' has n vertices and may have $m \geq 1$ components ($m > 1$ if v_{n+1} was a cut vertex). By the induction hypothesis, there is a spanning tree for each component of G' since each of these components has at most n vertices. Let $T_1, \ldots, T_m$ be the spanning trees for these components.

Let T' be the acyclic subgraph of G consisting of all the components' spanning trees. For each spanning tree, choose exactly one edge from E of the form $e^{(i)} = \{v_{n+1}, v^{(i)}\}$, where $v^{(i)}$ is a vertex in component i and add this edge to T' to create the tree T. (See Fig. 4.5.) It is easy to see that no cycle is created in T through these

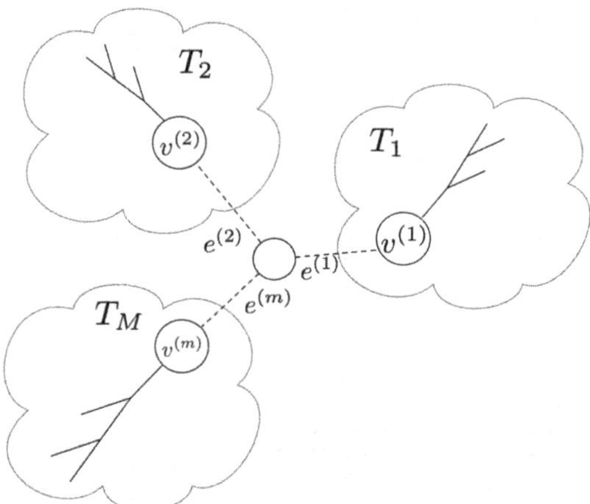

Fig. 4.5 Adding edges back in that were removed in the creation of the m components does not cause a cycle to form. The result is a spanning tree of the connected graph.

operations because, by construction, each edge $e^{(i)}$ is a cut edge, and by Corollary 3.44, it cannot lie on a cycle. The graph T contains every vertex of G and is connected and acyclic. Therefore, it is a spanning tree of G. The theorem follows by induction. □

Corollary 4.17. *Every graph $G = (V, E)$ has a spanning forest $F = (V, E')$.* □

Definition 4.18 (Leaf). Let $T = (V, E)$. If $v \in V$ and $\deg(v) = 1$, then v is called a *leaf* of T.

Lemma 4.19. *Every tree with one edge has at least two leaves.*

Proof. Let

$$w = (v_1, e_1, v_2, \ldots, v_n, e_n, v_{n+1})$$

be a path of maximal length in T. Consider vertex v_{n+1}. If $\deg(v_{n+1}) > 1$, then there are two possibilities, as follows: (i) There is an edge e_{n+1} and a vertex v_{n+2} with v_{n+2} *not* in the sequence w. In this case, we can extend w to w' defined as

$$w' = (v_1, e_1, v_2, \ldots, v_n, e_n, v_{n+1}, e_{n+1}, v_{n+2}),$$

which contradicts our assumption that w was maximal in length. (ii) There is an edge e_{n+1} and a vertex v_{n+2} and for some $k \in \{1, \ldots, n\}$, $v_{n+2} = v_k$; i.e., v_{n+2} is in the sequence w. In this case, there is a closed sub-walk

$$w' = (v_k, e_k, v_{k+1}, \ldots, v_{n+1}, e_{n+1}, v_{n+2}).$$

Since w is a path, there are no other repeated vertices in the sequence w'; thus, w' is a cycle in T, contradicting our assumption that T was a tree. The reasoning above holds for vertex v_1 as well; thus, the two end points of every maximal path in a nondegenerate tree must be leaves. This completes the proof. □

Corollary 4.20. *Let $G = (V, E)$ be a graph. If each vertex in V has a degree of at least 2, then G contains a cycle.* □

Lemma 4.21. *Let $T = (V, E)$ be a tree with $|V| = n$. Then, $|E| = n - 1$.*

Proof. We proceed by induction. For the case when $n = 1$, this statement must be true. Now, suppose that the statement is true for $|V| \leq n$. We show that when $|V| = n+1$, then $|E| = n$, assuming that $T = (V, E)$ is a tree. By Lemma 4.19, we know that if T is a tree, then it contains one component and at least two leaves. Therefore, choose a vertex $v \in V$ that is a leaf in T. There is some edge $e = \{v', v\} \in E$. Consider the graph $T' = (V', E')$ with $V' = V \setminus \{v\}$ and $E' = E \setminus \{e\}$. This new graph T' must:

(1) have one component since v was connected to only one other vertex $v' \in V$ and T had only one component; and
(2) be acyclic since T itself was acyclic and we have not introduced new edges to create a cycle.

Therefore, T' is a tree with n vertices, and by the induction hypothesis, it must contain $n - 1$ edges. Since we removed exactly one edge (and one vertex) to construct T' from T, it follows that T had exactly n edges and our originally assumed $n+1$ vertices. The required result follows immediately from induction. □

Corollary 4.22. *If $G = (V, E)$ is a forest with n vertices, then G has $n - c(G)$ edges. (Recall that $c(G)$ is the number of components in G.)* □

Theorem 4.23. *A graph $G = (V, E)$ is connected if and only if it has a spanning tree.* □

Theorem 4.24 (Tree characterization theorem). *Let $T = (V, E)$ be a graph with $|V| = n$. Then, the following are equivalent:*

(1) *T is a tree.*
(2) *T is acyclic and has exactly $n - 1$ edges.*
(3) *T is connected and has exactly $n - 1$ edges.*
(4) *T is connected and every edge is a cut edge.*
(5) *Any two vertices of T are connected by exactly one path.*
(6) *T is acyclic and the addition of any new edge creates exactly one cycle in the resulting graph.*

Proof. ($1 \implies 2$) Assume T is a tree. Then, by definition, T is acyclic and the fact that it has $n - 1$ edges follows from Lemma 4.21.

($2 \implies 3$) Since T is acyclic, it must be a forest, and by Corollary 4.22, $|E| = n - c(T)$. Since we assumed that T has $n-1$ edges, we must have $n - c(T) = n - 1$, thus the number of components of T is 1 and T must be connected.

($3 \implies 4$) The fact that T is connected is assumed from (3). Suppose we consider the graph $T' = (V, E')$, where $E' = E \setminus \{e\}$. Then, the number of edges in T' is $n-2$. The graph T' contains n vertices and must still be acyclic (that is a forest); therefore, $n-2 = n - c(T')$. Thus, $c(T') = 2$ and e is a cut edge.

($4 \implies 5$) Choose two vertices v and v' in V. The fact that there is a path between v and v' is guaranteed by our assumption that T is connected. By way of contradiction, suppose that there are at least two paths from v to v' in T. These two paths must diverge at some vertex $w \in V$ and recombine at some other vertex w'. (See Fig. 4.6.) We can construct a cycle in T by beginning at vertex w, following the first path to w', and then following the second path back to w from w'.

By Theorem 3.43, removing any edge in this cycle cannot result in a disconnected graph. Thus, no edge in the constructed cycle is a cut edge, contradicting our assumption about T. Thus, two paths connecting v and v' cannot exist.

($5 \implies 6$) The fact that any pair of vertices is connected in T implies that T is connected (i.e., it has one component). Now, suppose that T has a cycle (like the one illustrated in Fig. 4.6). Then, it is easy to see that there are (at least) two paths connecting w and w', contradicting our assumption. Therefore, T is acyclic. The fact that adding an edge creates exactly one cycle can be seen in the following way: Consider two vertices v and v' and suppose the edge $\{v, v'\}$ is not in E. We know that there is a path

$$(v, \{v, u_1\}, u_1, \ldots, u_n, \{u_n, v'\}, v')$$

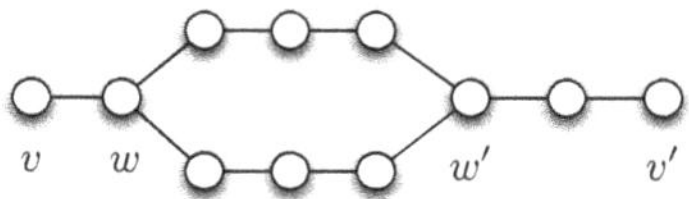

Fig. 4.6 The proof of $4 \implies 5$ requires us to assume the existence of two paths in graph T connecting vertex v to vertex v'. This assumption implies the existence of a cycle, contradicting our assumptions about T.

in T connecting v and v' and it is unique. Adding the edge $\{v, v'\}$ creates the cycle

$$c_1 = (v, \{v, u_1\}, u_1, \ldots, u_n, \{u_n, v'\}, v', \{v, v'\}, v),$$

so at least one cycle is created. To see that this cycle is unique, note that if there is another cycle present, then it must contain the edge $\{v, v'\}$. Suppose that this cycle is

$$c_2 = (v, \{v, w_1\}, w_1, \ldots, w_n, \{w_n, v'\}, v', \{v, v'\}, v),$$

where there is at least one vertex w_i not present in the set $\{u_1, \ldots, u_n\}$ (otherwise, the two cycles are identical). We now see that there must be two disjoint paths connecting v and v', namely

$$(v, \{v, u_1\}, u_1, \ldots, u_n, \{u_n, v'\}, v')$$

and

$$(v, \{v, w_1\}, w_1, \ldots, w_n, \{w_n, v'\}, v').$$

This contradicts our assumption about T. Thus, the created cycle is unique.

($6 \implies 1$) It suffices to show that T has a single component. Suppose that it is not so; there are at least two components of T. Choose two vertices v and v' in V so that these two vertices are not in the same component. Then, the edge $e = \{v, v'\}$ is not in E, and adding it to E cannot create a cycle. To see why, note that if T' is the graph that results from the addition of e, then e is now a cut edge. Applying Corollary 3.44, we see that e cannot lie on a cycle; thus, the addition of this edge does not create a cycle, contradicting our assumption about T. Thus, T must have a single component. Since it is acyclic and connected, T is a tree. This completes the proof. □

Definition 4.25 (Tree-graphic sequence). Recall that from Definition 2.13, a tuple $\mathbf{d} = (d_1, \ldots, d_n)$ is graphic if there exists a graph G with a degree sequence of $\mathbf{d}$. The tuple $\mathbf{d}$ is *tree-graphic* if it is both graphic *and* there exists a tree with a degree sequence of $\mathbf{d}$.

Theorem 4.26. *A degree sequence* $\mathbf{d} = (d_1, \ldots, d_n)$ *is tree-graphic if and only if*

(1) $n = 1$ $d_1 = 0$;
(2) $n \geq 2$, $d_i > 0$ *for* $i = 1, \ldots, n$; *and*

$$\sum_{i=1}^{n} d_i = 2n - 2. \tag{4.5}$$

Remark 4.27. One direction of the proof is left as an exercise.

Proof. ($\Leftarrow$) Now, suppose that Eq. (4.5) holds. If $n = 1$, then $d_1 = 0$, and this is a degenerate tree (with one vertex). We now proceed by induction to establish the remainder of the theorem. If $n = 2$, $2n - 2 = 2$, and if $d_1, d_2 > 0$, then $d_1 = d_2 = 1$ by necessity. This is the degree sequence for a tree with two vertices joined by a single edge; thus, it is a tree-graphic degree sequence. Now, assume the statement holds for all integers up to some n. We show that it is true for $n + 1$. Consider a degree sequence $(d_1, \ldots, d_{n+1})$ such that

$$\sum_{i=1}^{n+1} d_i = 2(n+1) - 2 = 2n.$$

We assume that the degrees are ordered (largest first) and positive. Therefore, $d_1 \geq 2$ (because otherwise, $d_1 + \cdots + d_{n+1} \leq n+1$). We also assume that $d_1 \leq n$. Note that in the case where $d_1 = n$, we must have $d_2 = d_3 = \cdots = d_{n+1} = 1$. Moreover, if $d_1 = d_2 = \cdots = d_{n-1} = 2$, then $d_n = d_{n+1} = 1$. Since $d_i > 0$ for $i = 1, \ldots, n+1$ from the previous two facts, we see that for any positive value of d_1, we must have $d_n = d_{n+1} = 1$ in order to ensure that $d_1 + d_2 + \cdots + d_{n+1} = 2n$. Consider the sequence of degrees

$$\mathbf{d}' = (d_1 - 1, d_2, \ldots, d_n).$$

Since $d_{n+1} = 1$, we can see that $(d_1 - 1) + d_2 + \cdots + d_n = 2n - 2$. Thus, a permutation of $\mathbf{d}'$ to correct the order leads to a tree-graphic sequence by the induction hypothesis. Let T' be the tree that results from this tree-graphic degree sequence, and let v_1 be the vertex with a degree of $d_1 - 1$ in T'. Then, by adding a new vertex v_{n+1} to T' along with edge $\{v_1, v_{n+1}\}$, we have constructed a tree T with the original degree sequence of $\mathbf{d}$. This new graph T must be connected since T' was connected, and we have connected v_{n+1} to a vertex in T'. The new graph must be acyclic since v_{n+1} does not appear in T'. The result follows by induction. $\square$

4.3 Eulerian Graphs

Remark 4.28. Suppose we have the walk

$$w = (v_1, e_1, v_2, \ldots, v_n, e_n, v_{n+1}).$$

If for some $m \in \{1, \ldots, n\}$ and for some $k \in \mathbb{Z}$, we have $v_m = v_{m+k}$. Then,

$$w' = (v_m, e_m, \ldots, e_{m+k-1}, v_{m+k})$$

is a *closed sub-walk* of w. The walk w' can be deleted from the walk w to obtain a new walk

$$w'' = (v_1, e_1, v_2, \ldots, v_{m+k}, e_{m+k}, v_{m+k+1}, \ldots, v_n, e_n, v_{n+1})$$

that is shorter than the original walk. This is illustrated in Fig. 4.7.

Lemma 4.29. *Let $G = (V, E)$ be a graph, and suppose that t is a nontrivial tour (closed trail) in G. Then, t contains a cycle.*

Proof. The fact that t is closed implies that it contains at least one pair of repeated vertices. Therefore, a closed sub-walk of t must exist since t itself has these repeated vertices. Let c be a minimal (length) closed sub-walk of t. We show that c must be a cycle. By way of contradiction, suppose that c is not a cycle. Then, since it is closed, it must contain a repeated vertex (that is not its first vertex). If we applied our observation from Remark 4.28, we could produce a smaller closed walk c', contradicting our assumption that c was minimal. Thus, c must have been a cycle. This completes the proof. □

Theorem 4.30. *Let $G = (V, E)$ be a graph and suppose that t is a nontrivial tour (closed trail). Then, t is composed of edge-disjoint cycles.*

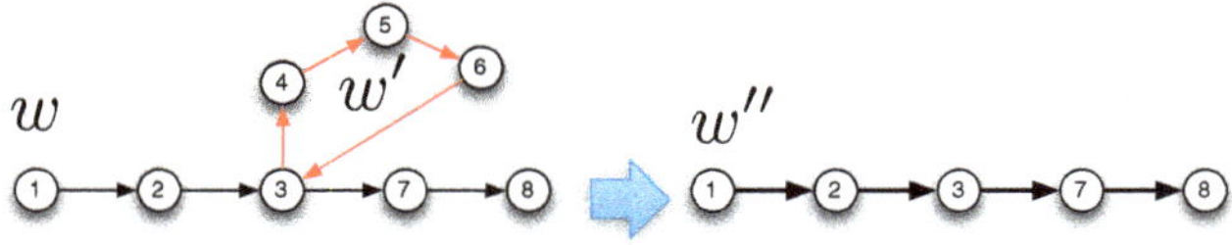

Fig. 4.7 We can create a new walk from an existing walk by removing closed sub-walks from the walk.

Proof. We proceed by induction on the length of the tour. In the base case, assume that t is a one-edge closed tour, then G is a non-simple graph that contains a self-loop, and this is a single edge in t; thus, t is a (non-simple) cycle.[1] Now, suppose the theorem holds for all closed trails of length N or less. We show that the result holds for a tour of length $N+1$. Applying Lemma 4.29, we know that there is at least one cycle c in t. If we remove c from this tour, then we obtain a new tour t' (containing all the remaining edges). We note some vertices may be disconnected. These can be ignored. The tour t' has a length of at most N. We can now apply the induction hypothesis to see that this new tour t' is composed of disjoint cycles. From this construction, we conclude that t is composed of edge-disjoint cycles. The theorem is illustrated in Fig. 4.8. This completes the proof. □

Remark 4.31. This final theorem completely characterizes Eulerian graphs. We use results derived from our study of trees to prove the following theorem.

Theorem 4.32 (Eulerian graph characterization theorem). *Let $G = (V, E)$ be a nonempty, nontrivial connected graph G. Then, the following are equivalent:*

(1) *G is Eulerian.*
(2) *The degree of every vertex in G is even.*

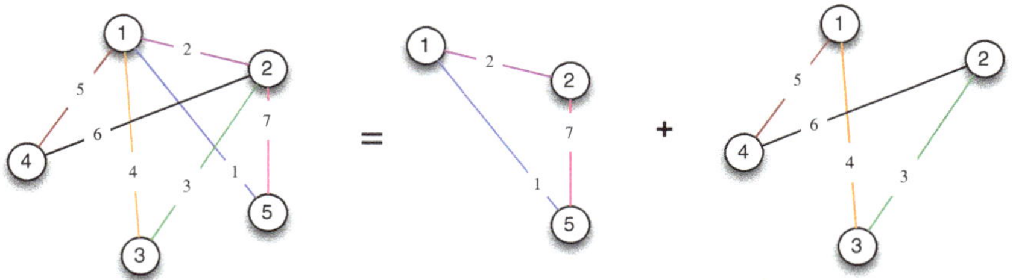

Fig. 4.8 We show how to decompose a (Eulerian) tour into an edge-disjoint set of cycles, thus illustrating Theorem 4.30.

[1] If we assume that G is simple, then the base case begins with t having a length of 3. In this case, it is a 3-cycle.

(3) *The set E is the union of the edge sets of a collection of edge-disjoint cycles in G.*

Proof. ($1 \implies 2$) Assume G is Eulerian, then there is an Eulerian tour t of G. Let v be a vertex in G. Each time v is traversed while following the tour, we must enter v by one edge and leave by another. Thus, v must have an even number of edges adjacent to it. If v is the initial (and final) vertex in the tour, then we leave v in the very first step of the tour and return in the last step; thus, the initial (and final) vertex of the tour must also have an even degree. Thus, every vertex has an even degree.

($2 \implies 3$) Since G is connected and every vertex has an even degree, it follows that the degree of each vertex is at least 2. Applying Corollary 4.20, we see that G must contain a cycle C. If this cycle includes every edge in G, then (3) is established. Suppose otherwise. Consider the graph G' obtained by removing all edges in C. If we consider C as a subgraph of G, then each vertex in C has exactly two edges adjacent to it. Thus, if v is a vertex in the cycle, then removing those edges in C that are adjacent to it will result in a vertex v having two fewer edges in G' than it did in G. Since we assumed that every vertex in G had an even degree, it follows that every vertex in G' must also have an even degree since we removed either two or zero edges from each vertex in G to obtain G'. We can repeat the previous process of constructing a cycle in G' and, if necessary, forming G''. Since there are a finite number of edges in G, this process must stop at some point, and we will be left with a collection of edge-disjoint cycles $\mathcal{C} = \{C, C', \ldots\}$ whose union is the entire edge set of G.

($3 \implies 1$) Assume that G is connected and that its edge set is the union of a collection of edge-disjoint cycles. We proceed by induction on the number of cycles. If there is only one cycle, then we simply follow this cycle in either direction to obtain a tour of G. Now, suppose that the statement is true for a graph whose edge set is the union of $\leq n$ edge-disjoint cycles. We'll show that the statement is true for a graph whose edge set is composed of $n + 1$ edge-disjoint cycles. Denote the cycles as $C_1, \ldots, C_{n+1}$. A subgraph G' of G composed of only cycles $C_1, \ldots, C_n$ will have m components, with $1 \leq m \leq n$. Each component is composed of at most n edge-disjoint cycles; therefore, applying the induction hypothesis, each has a tour. Denote the components as $K_1, \ldots, K_m$. The cycle C_{n+1} shares one vertex in common with at least one of these components (and perhaps all of them). Without loss of generality, assume that K_1 is a component sharing a vertex in common with C_{n+1} (if not, reorder the components to make this true). Begin following the tour around K_1 until we encounter the vertex v_1 that component K_1 and C_{n+1} share. At this point, break the tour of K_1 and begin traversing C_{n+1} until (i) we return to v_1 or (ii) we encounter a vertex v_2 that is shared by another component (say K_2). In case (i), we complete the tour of K_1, and necessarily, we must have completed a tour of the entire graph since it is connected. In case (ii), we follow the tour of K_2 until we return to v_2 and then continue following C_{n+1} until either case (i) occurs or case (ii) occurs again. In either case, we apply the same logic as before. Since there are a finite number of components, this process will eventually terminate with case (i), we complete the tour of K_1, and thus, we would have constructed a tour of the entire graph. Figure 4.8 also serves to illustrate this theorem. □

4.4 Chapter Notes

Bipartite graphs appear frequently in applications, and we will see them again in Chapter 6, when we study matching problems. They occur in optimization in assignment problems (such as assigning jobs to workers) and in transportation problems, in which the flow of commodities from warehouses to stores is optimized. Reference [29] has a complete introduction to this subject. Bipartite graphs also appear in coding theory [32]. They are also used to describe Petri nets [33], which can be used to define simple discrete dynamical systems. Bipartite graphs are particularly useful in social network analysis [34] for modeling individuals and their attributes. For example, the bipartite graph between users and the websites they visit may be mined to obtain socially relevant information.

Trees are ubiquitous in computer science. They appear in several algorithms that are fundamental to combinatorial optimization, as we'll discuss in Chapter 5. Reference [28] contains an extensive introduction to algorithms, many of which use trees. Knuth also provides extensive information on trees and computer science [35] in his extensive work on programming. Family trees are a popular application of trees in everyday use.

Eulerian graphs and Eulerian trails/tours are used in bioinformatics [36] as well as CMOS circuit design [37]. Recently, Eulerian tours have also been used as a mechanism for displaying information [38]. Chapter 17 of Godsil and Royle's work [39] discusses the interesting relationship between knots (mathematical abstractions of shoelace knots) and Eulerian tours. Also, the classic route inspection problem (also known as the Chinese postman problem) has, at its core, the problem of finding an optimal Eulerian tour [40].

4.5 Exercises

Exercise 4.1
A graph has a cycle with a length of 15. Is it bipartite? Why?

Exercise 4.2
Prove Proposition 4.12. [Hint: Use Theorem 4.8.]

Exercise 4.3
Prove Corollary 4.17.

Exercise 4.4
Prove Corollary 4.20.

Exercise 4.5
Prove Corollary 4.22.

Exercise 4.6
Prove Theorem 4.23.

Exercise 4.7
Decide whether the following degree sequence is tree graphic: $(3, 2, 2, 1, 1, 1)$. If it is, draw a tree with this degree sequence.

Exercise 4.8
Draw a graph that is Eulerian but Hamiltonian cycle. Explain the reason for your answer.

Exercise 4.9
The following sequence is graphic $(6, 2, 2, 2, 2, 2, 2, 2, 2)$. Is this graph Eulerian? Why?

Exercise 4.10
Prove the necessity part ($\Rightarrow$) of Theorem 4.26. [Hint: Use Theorem 2.10.]

Exercise 4.11
Show by example that Theorem 4.32 does not necessarily hold if we are only interested in Eulerian *trails*.

Part 2

Optimization in Graphs and NP-Completeness

Chapter 5

Trees, Algorithms, and Matroids

Remark 5.1 (Chapter goals). We study algorithms in this chapter. We focus on graph search algorithms in the form of breadth- and depth-first searches. We then discuss minimum spanning trees. Algorithms for finding minimum distance paths in graphs are then considered. We show an interesting application to currency trading that uses the Floyd–Warshall algorithm. The chapter concludes with a more theoretical treatment of matroids.

5.1 Two-Tree Search Algorithms

Remark 5.2. For the remainder of this section, we use the following, rather informal, definition of an algorithm: An *algorithm* is a set of steps (or operations) that can be followed to achieve a certain goal. We can think of an algorithm as having an input x and from which we obtain an output y. See Ref. [41] for a formal definition of algorithms in terms of Turing machines.

Remark 5.3. Tree searching is the process of enumerating the vertices of a tree (for the purpose of "searching" them). One can consider this process as generating a walk containing all the vertices of the tree at least once or as a way to create a sequence of the vertices. In this section, we take the latter view, though it will be clear how to create a walk as well.

The first algorithm, called breadth-first search (BFS), explores the vertices of a tree starting from a given vertex by exploring all the neighbors of this given vertex, then the neighbors of the neighbors, and so on until all the vertices have been encountered. The algorithm in pseudocode is shown in Algorithm 1.

```
Breadth-First Search on a Tree
Input: T = (V, E) a tree, v0 a starting vertex
Initialize: F = (v0) {a sequence of vertices to enumerate}
Initialize: F_next = () {the sequence of next vertices to enumerate}
Initialize: w = () {the sequence of vertices traversed}

while F ≠ ∅
    for each v ∈ F do
        Remove v from F
        Append v to w
        for each v' ∈ N(v) do
            if v' ∉ w then
                Append v' to F_next
            end if
        end for
    end for
    F = F_next
    F_next = ()
end while

Output: w {a breadth-first sequence of vertices in T}
```

Algorithm 1: Breadth-first search.

Example 5.4. Figure 5.1 shows the order in which the vertices are added to w during a BFS of the tree. We start at Vertex a as Vertex 1 and explore its immediate neighbors, adding Vertices b and c (2 and 3) to our list of vertices. From there, we explore the neighbors of Vertices b and c to obtain Vertices d and e as the fourth and fifth vertices enumerated.

Proposition 5.5. *A BFS of a tree* $T = (V, E)$ *enumerates all vertices.*

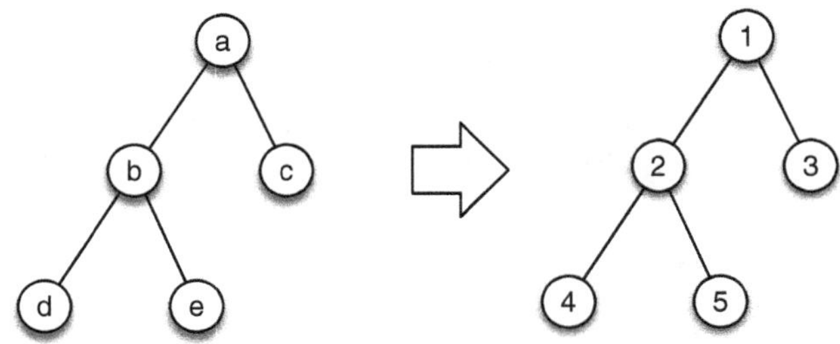

Fig. 5.1 The breadth-first walk of a tree explores the tree in an ever-widening pattern.

Proof. We proceed by induction. If T has one vertex, then clearly v_0 in the algorithm is that vertex. The vertex is added to w in the first iteration of the while loop at Line 1 and F_{next} is the empty set, thus the algorithm terminates. Now, suppose that the statement is true for all trees with at most n vertices. We show that the statement is true for a tree with $n+1$ vertices. To see this, construct a new tree T' in which we remove a leaf vertex v' from T. Clearly, the algorithm must enumerate every vertex in T'; therefore, there is a point in which we reach Line 3 with some vertex v that is adjacent to v' in T. At this point, v' would be added to F_{next}, and it would be added to w in the next execution through the while loop since $F \neq \emptyset$ the next time. Thus, every vertex of T must be enumerated in w. This completes the proof. □

Remark 5.6. BFS can be modified for directed trees in an obvious way. Necessarily, we need v_0 to be strongly connected to every other vertex in order to ensure that BFS enumerates every possible vertex.

Remark 5.7. Another algorithm for enumerating the vertices of a tree is the depth-first search (DFS) algorithm. This algorithm works by descending into the tree as deeply as possible (until a leaf is identified) and then working back up. We present DFS as a *recursive* algorithm in Algorithm 2. A *recursive algorithm* is an algorithm that calls itself during its own execution.

```
Depth-First Search on a Tree
Input: T = (V, E) a tree, v_0 a starting vertex
Initialize: v_now = v_0 {the current vertex}
Initialize: w = (v_0) {the sequence of next vertices to enumerate}

Recurse(T, v_now, w)

Output: w {a depth-first sequence of vertices in T}

Recurse
Input: T = (V, E) a tree, v_now current vertex, w the sequence

for each v ∈ N(v_now) do
    if v ∉ w then
        Append v to w
        Recurse(T, v, w)
    end if
end for
```

Algorithm 2: Depth-first search.

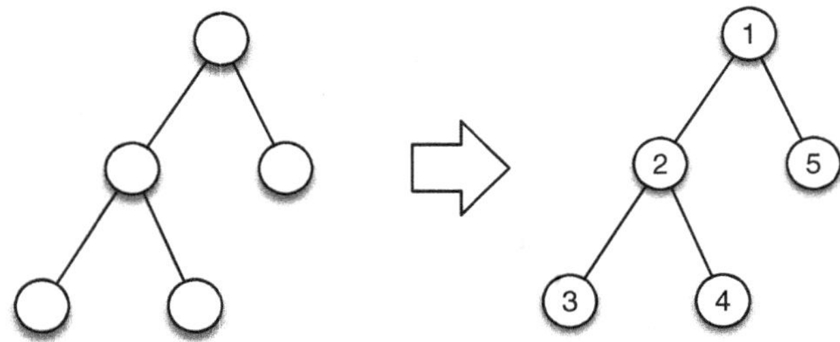

Fig. 5.2 The depth-first walk of a tree explores the tree in an ever-deepening pattern.

Example 5.8. Figure 5.2 shows the order the vertices are added to w during a DFS of the tree. We start at Vertex 1 and then explore down to Vertex 3 before stopping because Vertex 3 is a leaf. We then go back to Vertex 2 and explore from there down to Vertex 4 (which is a leaf). We then return back to Vertex 1 and explore down from there to Vertex 5.

Proposition 5.9. *A DFS of a tree $T = (V, E)$ enumerates all vertices in w.* □

Remark 5.10. We note that BFS and DFS can be trivially modified to search through connected graph structures and construct spanning trees for these graphs. We also note that BFS and DFS can be modified to function on directed trees (and graphs) and that all vertices will be enumerated provided that every vertex is reachable by a directed path from v_0.

Remark 5.11. In terms of implementation, we note that the recursive implementation of DFS works on most computing systems, provided the graph in question has a longest path of at most some specified value. This is because most operating systems prevent a recursive algorithm from making any more than a specified number of recursion calls.

5.2 Building a BFS/DFS Spanning Tree

Remark 5.12. We can also build a spanning tree using a BFS or DFS on a graph. These algorithms are shown in Algorithms 3 and 4.

Note that instead of just appending vertices to w, we also grow a tree that will eventually span the input graphs G (just in case G is connected).

```
Breadth-First Search Spanning Tree
Input: G = (V, E) a graph, v_0 a starting vertex
Initialize: F = (v_0) {a sequence of vertices to enumerate}
Initialize: F_next = () {the sequence of next vertices to enumerate}
Initialize: w = () {the sequence of vertices traversed}
Initialize: T = (V, E') {the tree returned}

while F ≠ ∅
    for each v ∈ F do
        Remove v from F
        Append v to w
        for each v' ∈ N(v) do
            if v' ∉ w then
                Append v' to F_next
                Add {v, v'} to E'
            end if
        end for
    end for
    F = F_next
    F_next = ()
end while

Output: T {a breadth-first spanning tree of G}
```

Algorithm 3: Breadth-first search spanning tree.

```
Depth-First Search Spanning Tree
Input: G = (V, E) a graph, v_0 a starting vertex
Initialize: v_now = v_0 {the current vertex}
Initialize: w = (v_0) {the sequence of vertices enumerated}
Initialize: T = (V, E') {the tree to return}

Recurse(G, T, v_now, w)

Output: T {a depth-first spanning tree of G}

Recurse
Input: G = (V, E) a graph, T = (V, E') a tree, v_now current vertex, w the sequence

for each v ∈ N(v_now) do
    if v ∉ w then
        Append v to w
        Add {v_now, v} to E'
        Recurse(T, v, w)
    end if
end for
```

Algorithm 4: Depth-first search spanning tree.

Example 5.13. We illustrate the breadth-first and depth-first spanning tree constructions in Fig. 5.3. In the BFS, we start at a vertex

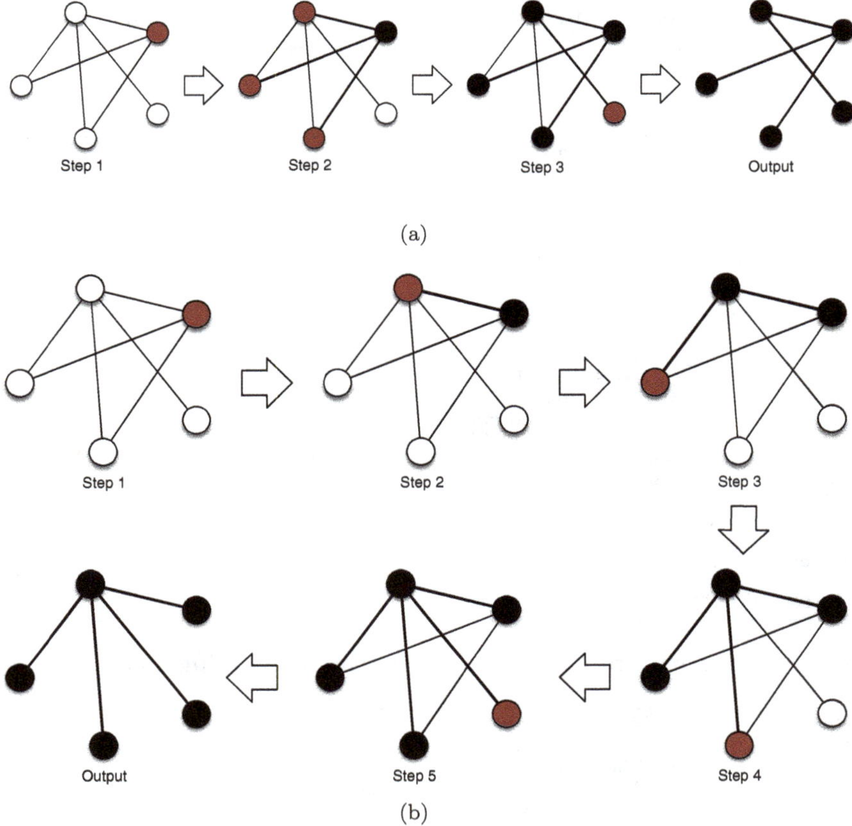

Fig. 5.3 (a) The construction of a breadth-first spanning tree is illustrated. (b) The construction of a depth-first spanning tree is illustrated.

and add it to a growing tree. We then add the vertices and edges that are adjacent to this graph, keeping track of the vertices just added. We repeat this process with the vertices just added, being careful *not* to repeat any vertices as this would create a cycle.

The DFS approach to building a spanning tree works similarly, except now we continue exploring along a path (adding vertices and edges as we go) until we can explore no further because we have hit a vertex with a degree of one or all the neighbors of that vertex have already been added to the tree. We then backtrack until we find a new direction to explore.

Note that the two approaches generate different spanning trees. This is not surprising since their search pattern is very different. Whether a DFS or BFS is used in exploring a graph depends on the situation.

Remark 5.14. Building a breadth- (or depth-) first spanning is a straightforward algorithmic way to check whether a graph is connected. In Chapter 6, we use a BFS as part of our analysis of flows on graphs.

5.3 Prim's Spanning Tree Algorithm

Definition 5.15 (Weighted graph). A *weighted graph* is a pair (G, w) where $G = (V, E)$ is a graph and $w : E \to \mathbb{R}$ is a weight function.

Remark 5.16. A weighted digraph is defined analogously.

Example 5.17. Consider the graph shown in Fig. 5.4. A weighted graph is simply a graph with a real number (the weight) assigned to each edge. Weighted graphs arise in several instances, such as travel planning and communications, as well as in graph-based data science.

Remark 5.18. Any graph can be thought of as a weighted graph in which we assign a weight of 1 to each edge. The distance between two vertices in a graph can then easily be generalized in a weighted graph. If $p = (v_1, e_1, v_2, \ldots, v_n, e_n, v_{n+1})$ is a path, then the weight

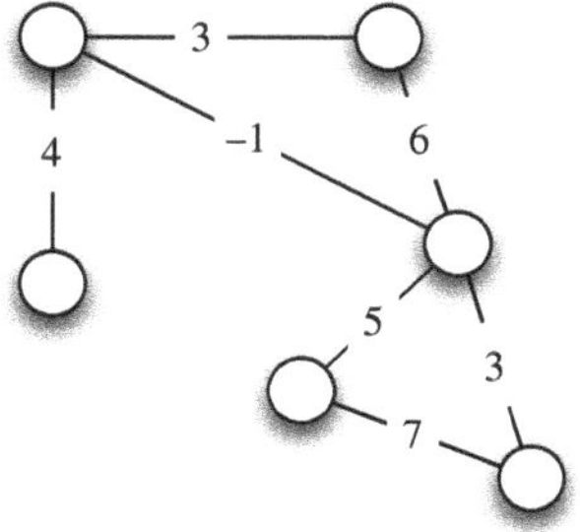

Fig. 5.4 A weighted graph is simply a graph with a real number (the weight) assigned to each edge.

of the path is

$$\sum_{i=1}^{n} w(e_i).$$

Thus, in a weighted graph, the distance between two vertices v_1 and v_2 is the weight of the least weight path connecting v_1 and v_2. We study the problem of finding this distance in Section 5.6.

Definition 5.19 (Graph weight). Let (G, w) be a weighted graph with $G = (V, E)$. If $H = (V', E')$ is a subgraph of G, then the *weight of H* is

$$w(H) = \sum_{e \in E'} w(e).$$

Definition 5.20 (Minimum spanning forest problem). Let (G, w) be a weighted graph with $G = (V, E)$. The *minimum spanning forest problem* for G is to find a forest $F = (V', E')$ that is a spanning subgraph of G that has the smallest possible weight.

Remark 5.21. If (G, w) is a weighted graph and G is connected, then the minimum spanning forest problem becomes the *minimum spanning tree* (MST) problem.

Example 5.22. An MST for the weighted graph shown in Fig. 5.4 is shown in Fig. 5.5. In the MST problem, we attempt to find a spanning subgraph of a graph G that is a tree and has minimal weight (among all spanning trees). We verify that the proposed spanning tree is minimal when we derive algorithms for constructing a minimum spanning forest.

Remark 5.23. The following algorithm, commonly called Prim's algorithm [42], constructs an MST for a connected graph.

Example 5.24. We illustrate the successive steps of Prim's algorithm in Fig. 5.6. At the start, we initialize our set $V' = \{1\}$ and the edge set is empty. At each successive iteration, we add an edge that connects a vertex in V' with a vertex not in V' that has minimum weight. Going from Iteration 1 to Iteration 2, we could have chosen to add either edge $\{1, 3\}$ or edge $\{4, 6\}$ (the order doesn't matter), so any tie-breaking method will suffice. We continue adding edges

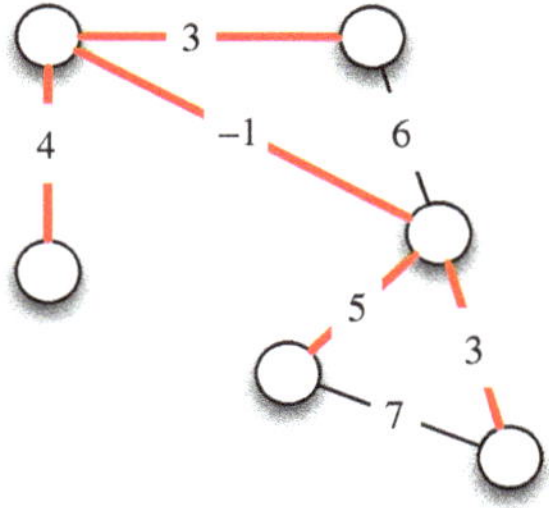

Fig. 5.5 In the minimum spanning tree problem, we attempt to find a spanning subgraph of a graph G that is a tree and has minimal weight (among all spanning trees).

Prim's Algorithm

Input: (G, w) a weighted connected graph with $G = (V, E)$, v_0 a starting vertex
Initialize: $E' = \emptyset$ {the edge set of the spanning tree}
Initialize: $V' = \{v_0\}$ {new vertices added to the spanning tree}

while $V' \neq V$
  Set $X := V \setminus V'$
  Choose edge $e = \{v, v'\}$ so (i) $v \in V'$; (ii) $v' \in X$ and:

$$w(e) = \min_{u \in U, u' \in X} w(\{u, u'\})$$

  Set $E' = E' \cup \{e\}$
  Set $V' = V' \cup \{v'\}$
end while

Output: $T = (V', E')$ {T is an MST.}

Algorithm 5: Prim's algorithm.

until all vertices in the original graph are in the spanning tree. Note that the output from Prim's algorithm matches the MST shown in Fig. 5.5.

Theorem 5.25. *Let (G, w) be a weighted connected graph. Then, Prim's algorithm returns an MST.*

Proof. We show by induction that at each iteration of Prim's algorithm, the tree (V', E') is a sub-tree of the MST T of (G, w). If this is the case, then at the termination of the algorithm, (V', E') must be equal to the MST T.

To establish the base case, note that at the first iteration, $V' = \{v_0\}$ and $E' = \emptyset$; therefore, (V', E') must be a sub-tree of T, which is an MST of (G, w). Now, suppose that the statement is true for all

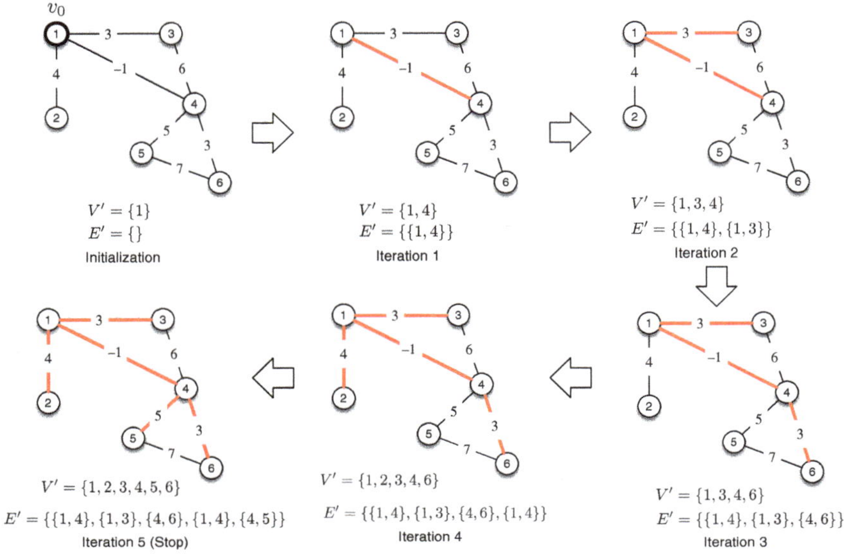

Fig. 5.6 Prim's algorithm constructs a minimum spanning tree by successively adding edges to an acyclic subgraph until every vertex is inside the spanning tree. Edges with minimal weight are added at each iteration.

iterations up to and including k, and let $T_k = (V', E')$ at iteration k. Suppose at iteration $k+1$, we add edge $e = \{v, v'\}$ to T_k to obtain $T_{k+1} = (U, F)$ with $U = V' \cup \{v'\}$ and $F = E' = E \cup \{e\}$. Suppose that T_{k+1} is not a sub-tree of T, then e is not an edge in T, and thus, e must generate a cycle in T. On this cycle, there is some edge $e' = \{u, u'\}$ with $u \in V'$ and $u' \notin V'$. At iteration $k+1$, we must have considered adding this edge to E'. We know by the selection of e that $w(e) \leq w(e')$. We have two cases:

Case 1: If $w(e) < w(e')$, then if we construct T' from T by removing e' and adding e, we know that T' must span G (this is illustrated in Fig. 5.7) and $w(T') < w(T)$; thus, T was not an MST of G, which contradicts our assumption about T. So, this case cannot occur.

Case 2: If $w(e) = w(e')$, then again constructing T' from T by removing e' and adding e, we obtain a second MST T', and T_{k+1} is a sub-tree of that MST.

Therefore, T_{k+1} must be a sub-tree of at least one MST. The result follows by induction. □

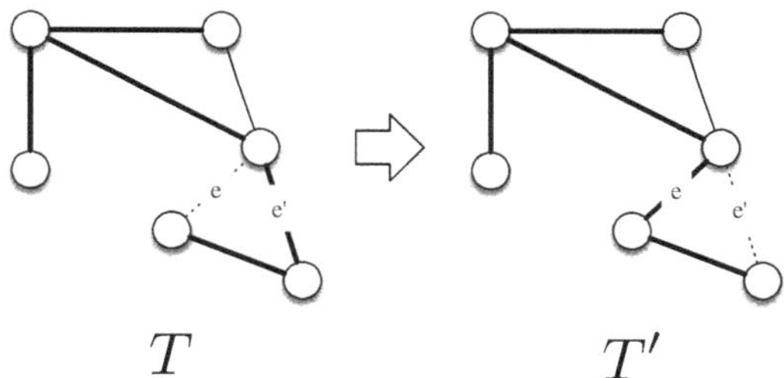

Fig. 5.7 When we remove an edge (e') from a spanning tree we disconnect the tree into two components. By adding a new edge (e) that connects vertices in these two distinct components, we reconnect the tree, and it is still a spanning tree.

5.4 Computational Complexity of Prim's Algorithm

Remark 5.26. In this section, we discuss computational complexity. This is a subject that has its own course in many computer science departments (and some math departments). Therefore, we can only scratch the surface of this fascinating topic, and we will *not* be able to provide completely formal definitions for all concepts. When definitions are informal, they will occur in remarks rather than in definition blocks.

Definition 5.27 (Big-O). Let $f, g : \mathbb{R} \to \mathbb{R}$. The function $f(x)$ is in the family $O(g(x))$ if there is an $N \in \mathbb{R}$ and a $c \in \mathbb{R}$ so that for all $x > N$, $|f(x)| \leq c|g(x)|$.

Remark 5.28. We have the following informal definition of *algorithmic running time*. The *running time* of an algorithm is the count of the number of steps an algorithm takes from the time we begin executing it to the time we obtain the output. We must be sure to include each time through any loops in the algorithm. This is not to be confused with the *wall clock* running time of an algorithm, which is dependent on the processor (a computer, a human, etc.) as well as the algorithmic details. Again, a more formal definition for algorithmic running time is given in Ref. [41].

Remark 5.29. In computing algorithmic running time, we need to be very careful about how we interpret the steps in the algorithm. For example, Prim's algorithm uses the word "Choose" in Line 3. But for a computer, this involves an enumeration of all the edges that might be connected to a specific vertex. Algorithm 6 does just that. In Lines 6–11, we enumerate over all possible edges and vertices that can be added to the spanning tree and find an edge with minimal weight and add it to the tree.

Prim's Algorithm (Explicit Form)

Input: (G, w) a weighted connected graph with $G = (V, E)$, v_0 a starting vertex
Initialize: $E' = \emptyset$ {the edge set of the spanning tree}
Initialize: $V' = \{v_0\}$ {new vertices added to the spanning tree}

(1) **while** $V' \neq V$
(2)   Set $X := V \setminus V'$
(3)   Set $e := \emptyset$
(4)   Set $w^* = \infty$
(5)   **for each** $v \in V'$
(6)   **for each** $v' \in X$
(7)   **if** $\{v, v'\} \in E$ **and** $w(\{v, v'\}) < w^*$
(8) $w^* = w(\{v, v'\})$
(9) $e := \{v, v'\}$
(10)   **end if**
(11)   **end for**
(12)   **end for**
(13)   Set $E' = E' \cup \{e\}$
(14)   Set $V' = V' \cup \{v'\}$
(15) **end while**

Output: $T = (V', E')$ {T is a minimum spanning tree.}

Algorithm 6: Prim's algorithm (explicit form).

Remark 5.30. Algorithm 6 is not optimal. It is intentionally not optimal so that we can compute its complexity in Theorem 5.31 easily and we do not have to appeal to special data structures. See Remark 5.32 for more on this.

Theorem 5.31. *The running time of Algorithm 6 is* $O(|V|^3)$.

Proof. Consider the steps in the `while` loop starting at Line 1. If there are n vertices, then at iteration k of the `while` loop, we know that $|V'| = k$ and $|X| = n - k$ since we add one new vertex to V' at each `while` loop iteration (and thus remove one vertex from X at each `while` loop iteration). The `for` loop beginning at Line 5 will have k iterations, and the `for` loop starting at Line 7 will have $n - k$ iterations. This means that for any iteration of the `while` loop, we perform $O(k(n - k))$ operations. Thus, for the whole algorithm, we perform

$$O\left(\sum_{k=1}^{n-1} k(n-k)\right) = O\left(\frac{1}{3}n^3 - \frac{1}{6}n\right)$$

operations. Thus, the running time for Algorithm 6 is $O(n^3) = O(|V|^3)$. □

Remark 5.32. As it turns out, the implementation of Prim's algorithm can have a *substantial* impact on the running time. There are implementations of Prim's algorithm that run in $O(|V|^2)$, $O(|E|\log(|V|))$, and $O(|E| + |V|\log(|V|))$ [28]. Thus, we cannot just say that the Prim's algorithm is an $O(g(x))$ algorithm; we must know which implementation of Prim's algorithm we are using. Clearly, the implementation in Algorithm 6 is not a very good one. Reference [28] has complete details on better implementations of Prim's algorithm using special ways of storing the data used in the algorithm.

Definition 5.33 (Polynomial running time). For a specific implementation of an algorithm, its running time is *polynomial* if there is some polynomial $p(x)$ so that when the running time of the algorithm is $f(x)$, then $f(x) \in O(p(x))$.

Theorem 5.34. *There is an implementation of Prim's algorithm that is polynomial.* □

5.5 Kruskal's Algorithm

Remark 5.35. In this section, we discuss Kruskal's algorithm [43], which is an alternative way to construct an MST of a weighted graph (G, w). The algorithm is shown in Algorithm 7.

Example 5.36. We illustrate Kruskal's algorithm in Fig. 5.8. The spanning subgraph starts with each vertex in the graph and no edges. In each iteration, we add the edge with the lowest edge weight, provided that it does not cause a cycle to emerge in the existing subgraph. It is purely coincidental that the construction of the spanning tree occurs in exactly the same set of steps as the Prim's algorithm. This is not always the case. In this example, edges that cause a cycle

Kruskal's Algorithm

Input: (G, w) a weighted connected graph with $G = (V, E)$ and $n = |V|$
Initialize: $Q = E$
Initialize: $V' = V$
Initialize: $E' = \emptyset$
Initialize: For all $v \in V$ define $C(v) := \{v\}$ $\{C(v)$ is the set of vertices with distance from v less than infinity at each iteration.$\}$

```
(1)  while |Q| > 0
(2)      Choose the edge e = (v, v') in Q with minimum weight.
(3)      if C(v) ≠ C(v')
(4)          for each u ∈ C(v): C(u) := C(u) ∪ C(v')
(5)          for each u ∈ C(v'): C(u) := C(u) ∪ C(v)
(6)          E' := E' ∪ {e}
(7)          Q := Q \ {e}
(8)      else
(9)          Q := Q \ {e}
(10)         GOTO 2
(11)     end if
(12) end while
```

Output: $T = (V', E')$ $\{T$ is a minimum spanning tree.$\}$

Algorithm 7: Kruskal's algorithm.

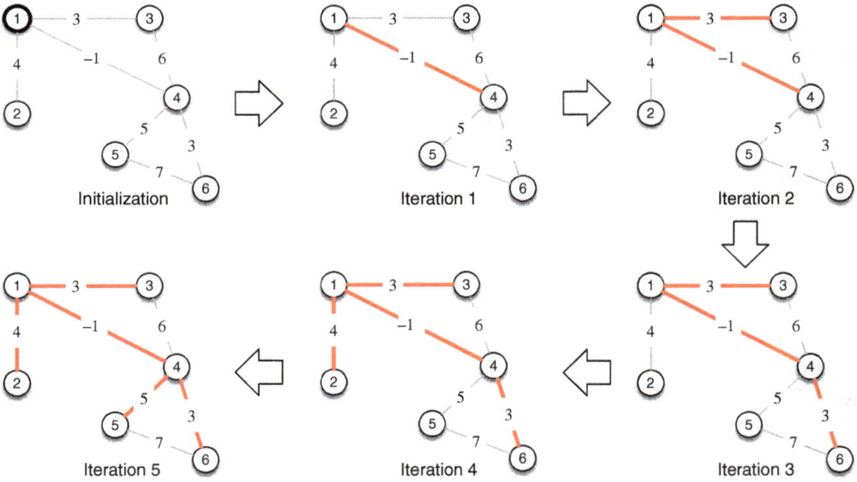

Fig. 5.8 Kruskal's algorithm constructs an MST by successively adding edges and maintaining an acyclic disconnected subgraph containing every vertex until that subgraph contains $n - 1$ edges, at which point we are sure it is a tree. Edges with minimal weight are added at each iteration.

are not chosen until the MST has been constructed. Also note that we have stopped the algorithm when the MST has been constructed. Strictly following Algorithm 7 would mean we should check the remaining edges (and thus remove them from Q).

Remark 5.37. We prove the following theorem in the last section of this chapter using a very generalized method. It can be shown by induction, just as we did in Theorem 5.25.

Theorem 5.38. *Let (G, w) be a weighted connected graph. Then, Kruskal's algorithm returns an MST.* □

Remark 5.39. Reference [28] contains a proof of the following theorem and shows how to use special data structures for the efficient implementation of Kruskal's algorithm.

Theorem 5.40. *There is an implementation of Kruskal's algorithm whose running time is $O(|E| \log(|V|))$.* □

5.6 Shortest Path Problem in a Positively Weighted Graph

Remark 5.41. The shortest path problem in a weighted graph is the problem of finding the least weight path connecting a given vertex v to a given vertex v'. Dijkstra's algorithm [44] answers this question by growing a spanning tree starting at v so that the unique path from v to any other vertex v' in the tree is the shortest. The algorithm is shown in Algorithm 8. It is worth noting that this algorithm *only* works when the weights in the graph are all positive. We discuss Floyd's algorithm [45], which solves this problem, later in the chapter.

Example 5.42. An example execution of Dijkstra's algorithm is shown in Fig. 5.9. At the start of the algorithm, we have Vertex 1 (v_0) as the vertex in the set Q that is closest to v_0 (it has a distance of 0, obviously). Investigating its neighbor set, we identify three vertices 2, 3, and 4, and the path length from Vertex 1 to each of these vertices is smaller than the initialized distance of ∞, so these vertices are assigned Vertex 1 as a parent, denoted by $p(v)$, and the new distances are recorded. Vertex 1 is then removed from the set Q.

Dijkstra's Algorithm

Input: (G, w) a weighted connected graph with $G = (V, E)$, v_0 an initial vertex.
Initialize: $Q := V$
Initialize: For all $v \in V$ if $v \neq v_0$ define $d(v_0, v) := \infty$ otherwise define $d(v_0, v) := 0$ $\{d(v_0, v)$ is the best distance from v_0 to $v.\}$
Initialize: For all $v \in V$, $p(v) :=$ **undefined** {A "parent" function that will be used to build the tree.}

while $Q \neq \emptyset$
  Choose $v \in Q$ so that $d(v_0, v)$ is minimal
  $Q := Q \setminus \{v\}$
  for each $v' \in N(v)$
    Define $\delta(v_0, v') := d(v_0, v) + w(\{v, v'\})$
    if $\delta(v_0, v') < d(v_0, v')$
      $d(v_0, v') := \delta(v_0, v')$
      $p(v') := v$
    end if
  end for
end while
Set $V' := V$
Set $E' := \emptyset$
for each $v \in V$
  if $p(v) \neq$ **undefined**
    $E' := E' \cup \{v, p(v)\}$
  end if
end for

Output: $T = (V', E')$ and $d(v_0, \cdot)$ $\{T$ is a Dijkstra tree, $d(v_0, \cdot)$ provides the distances.$\}$

Algorithm 8: Dijkstra's algorithm (adapted from Wikipedia's pseudocode, http://en.wikipedia.org/wiki/Dijkstra%27s_algorithm).

In the second iteration, we see that Vertex 3 is closest to v_0 (Vertex 1), and investigating its neighborhood, we see that the total distance from Vertex 1 to Vertex 3 and then from Vertex 3 to Vertex 4 is 9, which is smaller than the currently recorded distance of Vertex 1 to Vertex 4. Thus, we update the parent function of Vertex 4 so that it returns Vertex 3 instead of Vertex 1. We also update the distance function and continue to the next iteration. The next closest vertex to v_0 is Vertex 2. Investigating its neighbors shows that no changes need to be made to the distance or parent function. We continue in this way until all the vertices are exhausted.

Theorem 5.43. *Let (G, w) be a weighted connected graph with vertex v_0. Then, Dijkstra's algorithm returns a spanning tree T so that the distance from v_0 to any vertex v in T is the minimum distance from v_0 to v in (G, w).*

Proof. We proceed by induction to show that the distances from v_0 to every vertex $v \in V \setminus Q$ are correct when v is removed from Q.

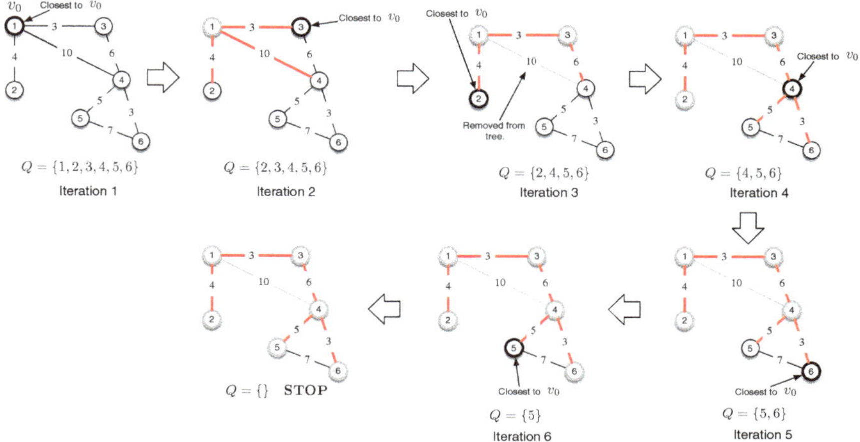

Fig. 5.9 Dijkstra's algorithm iteratively builds a tree of shortest paths from a given vertex v_0 in a graph. Dijkstra's algorithm can correct itself, as we see from Iterations 2 and 3.

To do this, define $X = V \setminus Q$, and let T_k be the tree generated by the vertices in X and the function $p(v)$.

In the base case, when v_0 is removed from Q (added to X), it is clear that $d(v_0, v_0) = 0$ is correct. Now, assume that the statement is true up to the kth iteration so that for any vertex v' added to X prior to the kth iteration, $d(v_0, v')$ is correct and the unique path in T_k from v_0 to v' defined by the function $p(v)$ is the minimum distance path from v_0 to v' in (G, w).

Before proceeding, note that for any vertex v' added to X at iteration k, $p(v')$ is fixed permanently after that iteration. Thus, the path from v_0 to v' in T_k is the same as the path from v_0 to v' in T_{k+1}. Therefore, assuming $d(v_0, v')$ and $p(v)$ to be correct at iteration k implies that they must be correct at all future iterations or, more generally, that they are correct for (G, w).

Suppose vertex v is added to set X (removed from Q) at iteration $k+1$, but the shortest path from v_0 to v is not the unique path from v_0 to v in the tree T_{k+1} constructed from the vertices in X and the function $p(v)$. Since G is connected, there is a shortest path, and we now have the following two possibilities: (i) The shortest path connecting v_0 to v passes through a vertex not in X; or (ii) the shortest path connecting v_0 to v passes through only vertices in X.

In the first case, if the true shortest path connecting v_0 to v passes through a vertex u not in X, then we have the following two new possibilities: (a) $d(v_0, u) = \infty$; or (b) $d(v_0, u) = r < \infty$. We may dismiss the first case as infeasible; thus, we have $d(v_0, u) = r < \infty$. In order for the distance from v_0 to v to be less along the path containing u, we know that $d(v_0, u) < d(v_0, v)$. But if that's true, then in Step 2 of Algorithm 8, we should have evaluated the neighborhood of u well before evaluating the neighborhood of v in the for loop starting at Line 4, and thus, u must be an element of X (i.e., not in Q). This contradicts our assumption and leads to the second case.

Suppose now that the true shortest path from v_0 to v leads to a vertex v'' before reaching v while the path recorded in T_{k+1} reaches v' before reaching v, as illustrated in the following figure.

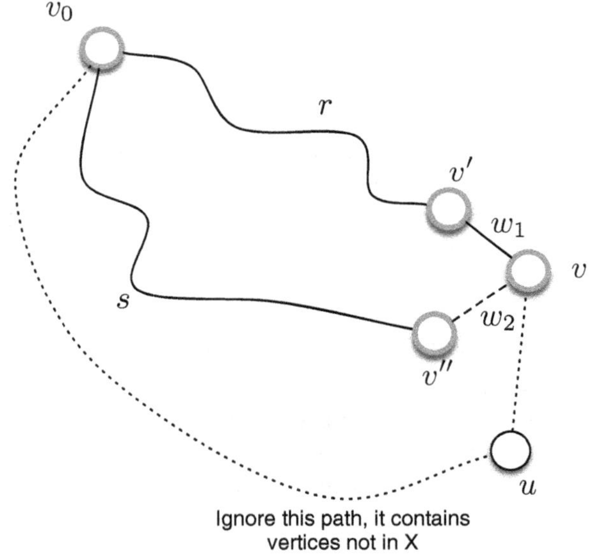

Let $w_1 = w(v', v)$ and $w_2 = w(v'', v)$. Then, it follows that $d(v_0, v') + w_1 > d(v_0, v'') + w_2$. By the induction hypothesis, $d(v_0, v')$ and $d(v_0, v'')$ are both correct as is their path in T_{k+1}. However, since both v' and v'' are in X, we know that the neighborhoods of both these vertices were evaluated in the `for` loop at Line 4. If when $N(v')$, we had $p(v) = v''$, then this would still be the case at iteration $k + 1$ since Line 6 specifically forbids changes to $p(v)$ unless

$d(v_0, v') + w_1 < d(v_0, v'') + w_2$. On the other hand, if $p(v) = v'$, when $N(v'')$ was evaluated, then it's clear at once that $p(v) = v''$ at the end of the evaluation of the `if` statement at Line 6 and would continue to be so at iteration $k + 1$. This contradicts our assumption on the structure of T_{k+1}. In either case, $d(v_0, v)$ could not be incorrect in T_{k+1} and the correctness of Dijkstra's algorithm follows from induction. □

Remark 5.44. A proof of the following theorem can be found in Ref. [28], which discusses special data structures for the implementation of Dijkstra's algorithm.

Theorem 5.45. *There is an implementation of Dijkstra's algorithm that has a running time of* $O(|E| + |V| \log(|V|))$. □

Remark 5.46. Dijkstra's algorithm is an example of a *dynamic programming* [46] approach to finding the shortest path in a graph. Dynamic programming is a sub-discipline of mathematical programming (or optimization), which we will encounter in the coming chapters.

5.7 Floyd–Warshall Algorithm

Remark 5.47. Dijkstra's algorithm is an efficient algorithm for graphs with non-negative weights. However, Dijkstra's algorithm can yield results that are incorrect when working with graphs with negative edge weights. To see this, consider the graph shown in Fig. 5.10.

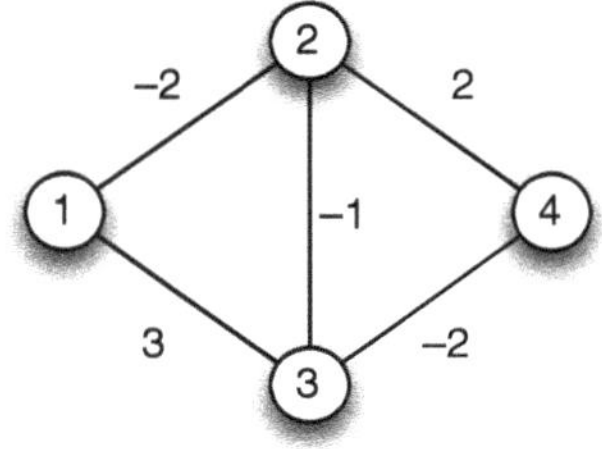

Fig. 5.10 This graph has negative edge weights that lead to confusion in Dijkstra's algorithm.

Executing Dijkstra's algorithm from Vertex 1 leads to the following data:

(1) At initialization, $Q = \{1, 2, 3, 4\}$ and $d(1, 1) = 0$ while $d(1, v) = \infty$ for $v \in \{2, 3, 4\}$. More importantly, $p(1) = p(2) = p(3) = p(4) =$ **undefined**. (Recall that $p(\cdot)$ is the *parent* function used to build the Dijkstra tree.)
(2) At the first step, Vertex 1 is removed from Q, and we examine its neighbors. During this step, we define: $d(1, 1) = 0$, $d(1, 2) = -2$, $d(1, 3) = 3$, and $d(1, 4) = \infty$ as well as $p(1) =$ **undefined**, $p(2) = p(3) = 1$, and $p(4) =$ **undefined**. Now, $Q = \{2, 3, 4\}$.
(3) At the second stage, Vertex 2 is the vertex closest to Vertex 1, so it is removed and we compute on its neighbors. We see that $d(1, 1) = -4$, $d(1, 2) = -2$, $d(1, 3) = -3$, and $d(1, 4) = 0$ as well as $p(1) = 2$, $p(2) = 1$, $p(3) = 2$, and $p(4) = 2$. Clearly, we have a problem already since $p(1) = 2$ and $p(2) = 1$, which means to get to vertex 1, we go through vertex 2 and vice versa. At this stage, $Q = \{3, 4\}$.
(4) We continue by removing Vertex 3 from Q and computing on its neighbors. We now have $d(1, 1) = -4$, $d(1, 2) = -4$, $d(1, 3) = -3$, and $d(1, 4) = -5$ as well as $p(1) = 2$ $p(2) = 3$, $p(3) = 2$, and $p(4) = 3$.
(5) Completing the algorithm and computing on the neighbors of 4 yields: $d(1, 1) = -4$, $d(1, 2) = -4$, $d(1, 3) = -7$, and $d(1, 4) = -5$ as well as $p(1) = 2$, $p(2) = 3$, $p(3) = 4$, and $p(4) = 3$. The resulting parent function cannot define a proper tree structure, and the algorithm fails.

These steps are illustrated in Fig. 5.11.

Remark 5.48. The real problem with Dijkstra's algorithm and negative edge weights is the fact that sequences of edges are repeating whose weights are negative. For example, going from Vertex 1 to Vertex 2 and back to Vertex 1 creates a lower-weight path than not leaving Vertex 1 at all. The result is a walk rather than a path. On a directed graph, this problem may not be as obvious, but the presence of a directed cycle with negative total length will cause problems. This is illustrated in Fig. 5.12. In these graphs, there is no shortest walk at all and the shortest length path (sometimes called a simple

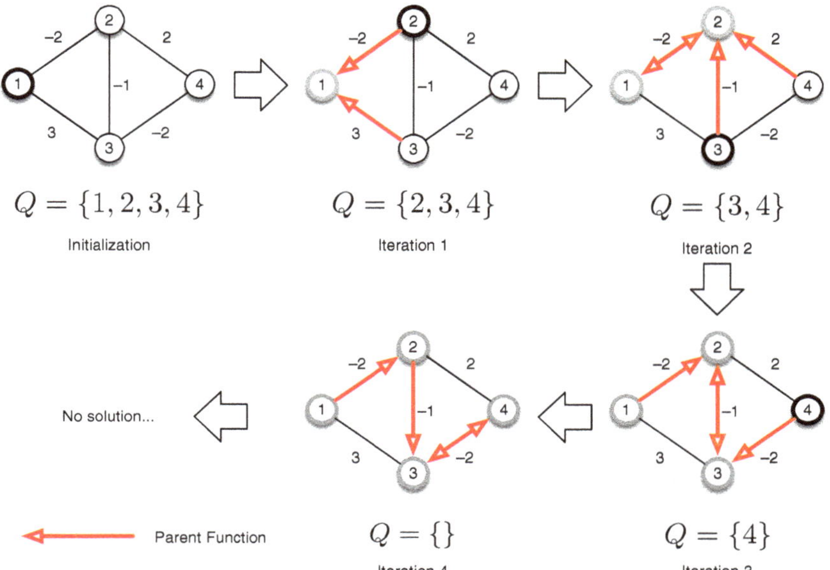

Fig. 5.11 The steps of Dijkstra's algorithm run on the graph in Fig. 5.10.

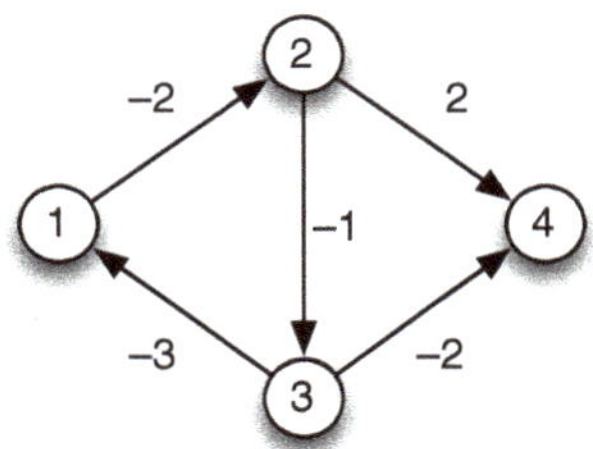

Fig. 5.12 A negative cycle in a (directed) graph implies that there is no shortest path between any two vertices, as repeatedly going around the cycle will make the path smaller and smaller.

path) can be very hard to find. (The problem is NP-hard, a type of problem we discuss in Chapter 7).

Remark 5.49. The problem of computing with negative edge weights can be solved through the Floyd–Warshall algorithm. This algorithm assumes a directed graph as input. The Floyd–Warshall algorithm for a directed graph is shown in Algorithm 9.

Floyd–Warshall Algorithm

Input: (G, w) a (directed) weighted connected graph with $G = (V, E)$, v_0 an initial vertex, v_f a final vertex

Initialize: For all $(u, v) \in V \times V$ if $e = (u, v) \in E$, then $d(u, v) := w(e)$; otherwise, if $u = v$, $d(u, v) := 0$ otherwise, $d(u, v) := \infty$. {Here, $d(u, v)$ is the shortest distance from u to v.}

Initialize: For all $(u, v) \in V \times V$, if $e = (u, v) \in E$, then $n(u, v) := v$; otherwise, $n(u, v) :=$ **undefined**. {The function $n(u, v)$ is the next vertex to move to when traversing from u to v along an optimal path.}

Assume: $V = \{v_1, \ldots, v_n\}$.

```
for each i ∈ {1, ..., n}
     for each u1 ∈ V
          for each u2 ∈ V
               if d(u1, vi) + d(vi, u2) < d(u1, u2)
                    d(u1, u2) := d(u1, vi) + d(vi, u2) {Update the distance.}
                    n(u1, u2) := n(u1, vi) {Update the next step to go through vi.}
          end for
     end for
end for
for each v ∈ V {Check for negative cycles.}
     if d(v, v) < 0
          RETURN NULL
     end if
end for
Set E' := ∅
Set V' := ∅
if n(v0, vf) ≠ undefined
     u := v0
     while u ≠ vf
          E' := E' ∪ (u, n(u, vf))
          V' := V' ∪ {u}
     end while
     V' = V' ∪ {vf} {Add the last step in the path.}
end if
```

Output: $P = (V', E')$ and $d(\cdot, \cdot)$ {P is a Floyd–Warshall path from v_0 to v_f, $d(\cdot, \cdot)$ provides the distances.}

Algorithm 9: Floyd–Warshall algorithm (adapted from Wikipedia's pseudocode, https://en.wikipedia.org/wiki/Floyd%E2%80%93Warshall_algorithm). This algorithm finds the shortest path between two vertices in a graph with (possibly) negative edge weights.

Remark 5.50. The Bellman–Ford algorithm is an older method that can also be used to solve this problem. It is less efficient than Dijkstra's algorithm. Johnson's algorithm combines the Bellman–Ford algorithm with Dijkstra's algorithm to solve problems with negative edge weights as well. The Bellman–Ford algorithm is used frequently in computer science because it has a variant that is distributed, called the distributed Bellman–Ford algorithm, which is used in developing computer networks.

Example 5.51. We illustrate the Floyd–Warshall algorithm on the graph shown in Fig. 5.13.

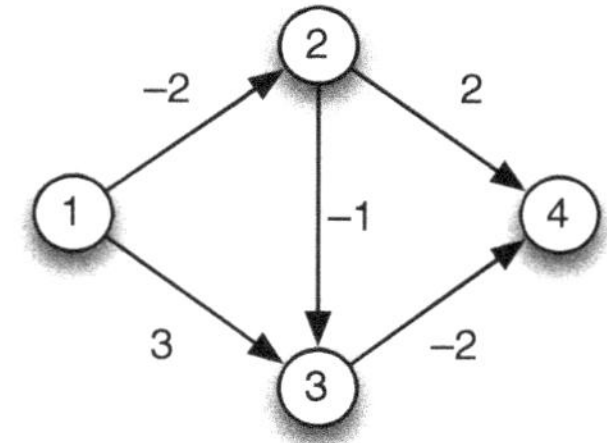

Fig. 5.13 A directed graph with negative edge weights.

Initially, the distance function is defined only for edges in the graph and is set to zero for the distance from a vertex to itself. Thus, we know that

(1) $d(v_k, v_k) = 0$ for $k = 1, 2, 3, 4$.
(2) $d(v_1, v_2) = -2$, $d(v_1, v_3) = 3$, $d(v_1, v_4) = \infty$.
(3) $d(v_2, v_1) = \infty$, $d(v_2, v_3) = -1$, $d(v_2, v_4) = 2$.
(4) $d(v_3, v_1) = \infty$, $d(v_3, v_2) = \infty$, $d(v_3, v_4) = -2$.
(5) $d(v_4, v_1) = \infty$, $d(v_4, v_2) = \infty$, $d(v_4, v_3) = \infty$.

There are four vertices in this example, so the outer loop will be executed four times. There will be a total of 64 comparisons at Line 4, and we cannot summarize them all. Instead, we discuss when the distance function changes.

Outer loop with v_1: In the outer loop with v_1, we are interested in paths that use v_1. Since v_1 has *no* in-edges, there are no paths that can be made shorter by passing through v_1. Thus, no change to the distance function is made.

Outer loop with v_2: In this loop, two things happen:

(1) When $u_1 = v_1$ and $u_2 = v_3$, the distance $d(v_1, v_3)$ is updated to -3 since there is a path of length -3 from v_1 *through* v_2 to v_3.
(2) When $u_1 = v_1$ and $u_2 = v_4$, the distance $d(v_1, v_4)$ is updated to 0 since there is a path of length 0 from v_1 *through* v_2 to v_4. (Before this, the distance from v_1 to v_4 was infinite.)

Outer loop with v_3: In this loop, two things happen:

(1) When $u_1 = v_1$ and $u_2 = v_4$, the distance from v_1 to v_4 is updated to -5 since there is a path of length -5 going through v_3 connecting v_1 to v_4.
(2) When $u_1 = v_2$ and $u_2 = v_4$, the distance from v_2 to v_4 is updated to -3.

Outer loop with v_4: In this loop, no further distance improvements can be made.

The complete distance function has the form:

(1) $d(v_k, v_k) = 0$ for $k = 1, 2, 3, 4$.
(2) $d(v_1, v_2) = -2$, $d(v_1, v_3) = -3$, $d(v_1, v_4) = -5$.
(3) $d(v_2, v_1) = \infty$, $d(v_2, v_3) = -1$, $d(v_2, v_4) = -3$.
(4) $d(v_3, v_1) = \infty$, $d(v_3, v_2) = \infty$, $d(v_3, v_4) = -2$.
(5) $d(v_4, v_1) = \infty$, $d(v_4, v_2) = \infty$, $d(v_4, v_3) = \infty$.

Theorem 5.52. *The Floyd–Warshall algorithm is correct. That is, the path returned connecting v_0 to v_f, the given initial and final vertices, is the shortest possible in the graph assuming no negative cycles exist.*

Proof. The proof is inductive on the outer for loop of the Floyd–Warshall algorithm. As in the algorithm statement, assume that we are provided a weighted directed graph (G, w), with $G = (V, E)$ and $V = \{v_1, \ldots, v_n\}$.

To execute this proof, we need an auxiliary function: Let u_1 and u_2 be vertices in the graph, and let $V_k = \{v_1, \ldots, v_k\} \subseteq V$. Let $d_k(u_1, u_2)$ be a function that returns the (shortest) distance between u_1 and u_2 using only the vertices in V_k as intermediary steps; that is, $d_k(u, v)$ is computed on the graph spanned by V_k, u_1, and u_2.

At the start of the algorithm (the base case), we have not executed the outermost `for` loop. For any pair of vertices u_1 and u_2, clearly $d_0(u_1, u_2)$ returns the shortest path considering only the vertices u_1 and u_2. Thus, $d_0(u_1, u_2)$ is equivalent to the function $d(\cdot, \cdot)$ in the Floyd–Warshall algorithm after initialization.

Now, assume that after k iterations of the outermost `for` loop, $d(\cdot, \cdot)$ defined in the Floyd–Warshall algorithm is identical to $d_k(\cdot, \cdot)$.

We show that after the $k+1$st iteration, $d(\cdot,\cdot)$ defined in the Floyd–Warshall algorithm is identical to $d_{k+1}(\cdot,\cdot)$. To see this, note that at Line 4, we determine whether $d(u_1, v_{k+1}) + d(v_{k+1}, u_2) < d(u_1, u_2)$; that is, we determine whether it is more expeditious to reach u_2 from u_1 via v_{k+1}. If not, we do nothing. If so, we update the function $d(\cdot,\cdot)$ to use this more expeditious path. Since $d(\cdot,\cdot)$ was equivalent to $d_k(\cdot,\cdot)$ by the induction hypothesis, it's clear that after the $k+$ 1st iteration of the outermost for loop, $d(\cdot,\cdot)$ must be equivalent to $d_{k+1}(\cdot,\cdot)$ by the construction that takes place at Line 5.

This induction must terminate after n steps, at which point we must have $d(\cdot,\cdot) = d_n(\cdot,\cdot)$. But this implies that the distances constructed from the Floyd–Warshall algorithm are correct since $d_n(\cdot,\cdot)$ is the true graph distance function. The construction of an optimal path from v_0 to v_f is ensured since $n(\cdot,\cdot)$ respects $d(\cdot,\cdot)$, which we just proved correct at the algorithm termination. □

Example 5.53 (Application of negative cycle detection). Suppose that we have n currencies with exchange rate $r_{i,j}$ when going from Currency i to Currency j. Imagine a scenario in which we start with \$1 of Currency 1, and for some $k \leq n$, we have $r_{1,2}r_{2,3}\cdots r_{k-1,k}r_{k,1} > 1$. Then, exchanging Currency 1 for Currency 2, Currency 2 for Currency 3, etc., will ultimately allow us to obtain more than \$1 of Currency 1. This is called *currency arbitrage.* If we assume that we have a digraph G with vertices $V = \{v_1, \ldots, v_n\}$ and with a directed edge from (v_i, v_j) for all pairs (i,j) with $i \neq j$, then the transformation from Currency 1 to Currency 2 to Currency 3... to Currency k and back to Currency 1 corresponds to a cycle in G. This is illustrated in Fig. 5.14. Let $w = (v_1, e_1, v_2, \ldots, v_n, e_n, v_{n+1})$ be a directed walk in the currency graph. Let r_w be the effective exchange rate from currency v_1 to v_{n+1} so that

$$r_w = r_{v_1,v_2} \cdot r_{v_2,v_3} \cdots r_{v_n,v_{n+1}}.$$

Note that

$$\log(r_w) = \log(r_{v_1,v_2}) + \log(r_{v_2,v_3}) + \cdots + \log(r_{v_n,v_{n+1}}). \tag{5.1}$$

Furthermore, if $r_{v_i,v_j} < 1$, then $\log(r_{v_i,v_j}) < 0$; this occurs when one unit of currency i is worth less than one unit of currency j. In finding currency arbitrage, we are interested in finding cycles with $r_w > 1$,

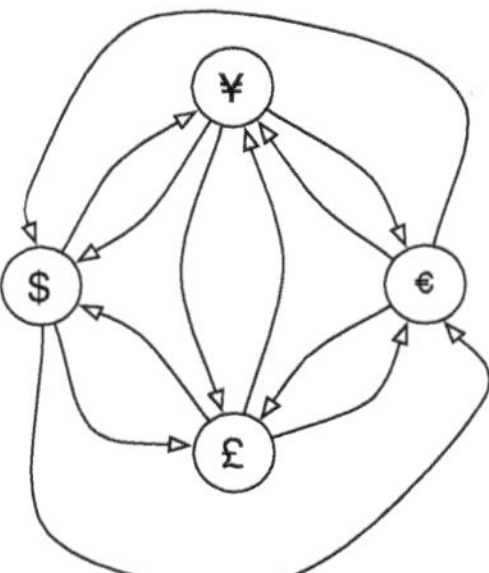

Fig. 5.14 A currency graph showing the possible exchanges. Cycles correspond to the process of going from one currency to another to another and ultimately ending up with the starting currency.

given an edge (v_i, v_j) of weight $-\log(r_{v_i,v_j})$. We can compute the exchange rate of a walk, path, or cycle by adding the negative log weights. Thus, if there is a cycle $c = (v_1, e_1, v_2, \ldots, v_n, e_n, v_1)$ with the property that $r_c > 1$, then there is a *negative weight cycle* in the currency graph with edge weights $-\log(r_{v_i,v_j})$. Thus, we can use the Floyd–Warshall algorithm to detect currency arbitrage.

5.8 Greedy Algorithms and Matroids

Definition 5.54 (Power set). Let S be a set, then 2^S is the *power set* of S; that is, 2^S is the set of all subsets of S.

Example 5.55. Consider the simple set $S = \{1, 2\}$. Then, the power set is

$$2^S = \{\emptyset, \{1\}, \{2\}, \{1, 2\}\}.$$

Here, $\emptyset$ is the *empty set*, and it is a subset of every set.

Definition 5.56 (Hereditary system). A *hereditary system* is a pair $(E, \mathcal{I})$ so that E is a finite set and $\mathcal{I} \subseteq 2^E$ a nonempty set of *independent sets* so that if $A \in \mathcal{I}$ and $B \subseteq A$, then $B \in \mathcal{I}$.

Remark 5.57. Any subset of E that is not in $\mathcal{I}$ is called a *dependent set.*

Proposition 5.58. *Suppose $G = (V, E)$ is a graph, and let $\mathcal{I}$ be the set of subsets of E such that if $E' \in \mathcal{I}$, then the subgraph of G induced by E' is acyclic (e.g., a sub-forest of G). Then, $(E, \mathcal{I})$ is a hereditary system.* □

Definition 5.59 (Weighted hereditary system). A *weighted hereditary system* is a triple $(E, \mathcal{I}, w)$ so that $(E, \mathcal{I})$ is a hereditary system and $w : E \to \mathbb{R}$ is a weight function on the elements of E.

Example 5.60. In light of Proposition 5.58, we could think of a weighted graph (G, w) with $G = (V, E)$ as giving rise to a weighted hereditary system $(E, \mathcal{I}, w)$ so that $\mathcal{I}$ is the collection of edge subsets of E that induce acyclic graphs and w is just the edge weighting.

Definition 5.61 (Minimum weight problem). Let $(E, \mathcal{I}, w)$ be a weighted hereditary system. Then, the *minimum weight problem* is to identify a set $E' \in \mathcal{I}$ (an independent set) such that

$$w(E') = \sum_{e \in E'} w(e) \tag{5.2}$$

is as small as possible and E' is a maximal subset of E (that is, there is no other set $I \in \mathcal{I}$ so that $E' \subset I$).

Remark 5.62. One can define a maximum weight problem in precisely the same way if we replace the word minimum with maximum and small with large. Algorithm 10 is called the greedy algorithm, and it can be used (in some cases) to solve the minimum weight problem. The name "greedy" makes a little more sense for maximum weight problems, but the examples we give are minimum weight problems.

Remark 5.63. Let (G, w) be a weighted graph and consider the weighted hereditary system with $(E, \mathcal{I})$, with $\mathcal{I}$ the collection of edge subsets of E that induce acyclic graphs and where w is just the edge weighting. Kruskal's algorithm is exactly a greedy algorithm. We begin with the complete set of edges and continue adding them to the forest (acyclic subgraph of a given weighted graph (G, w)), each time checking to make sure that the added edge does not induce a cycle (that is, that we have an element of $\mathcal{I}$). We use this fact shortly to prove Theorem 5.38.

```
Greedy Algorithm
Input: (E, I, w) a weighted hereditary system
Initialize: E' = ∅
Initialize: A = E

while A ≠ ∅
    Choose e ∈ A to minimize w(e)
    A := A \ {e}
    if E' ∪ {e} ∈ I
        E' := E' ∪ {e}
    end if
end while

Output: E'
```

Algorithm 10: Greedy algorithm (minimization).

Definition 5.64 (Matroid). Let $M = (E, \mathcal{I})$ be a hereditary system. Then, M is a *matroid* if it satisfies the *augmentation property*: If $I, J \in \mathcal{I}$ and $|I| < |J|$, then there is some $e \in E$ so that $e \in J$ and $e \notin I$ and so that $I \cup \{e\} \in \mathcal{I}$.

Remark 5.65. Definition 5.64 essentially says that if there are two independent sets (acyclic subgraphs are an example) and one has greater cardinality than the other, then there is some element (edge) that can be added to the independent set (acyclic graph) with smaller cardinality so that this new set is still independent (an acyclic graph).

Theorem 5.66. *Let $(E, \mathcal{I}, w)$ be a weighted hereditary system. The structure $M = (E, \mathcal{I})$ is a matroid if and only if the greedy algorithm solves the minimum weight problem associated with M.*

Proof. ($\Rightarrow$) Let $I = \{e_1, \ldots, e_n\}$ be the set in $\mathcal{I}$ identified by the greedy algorithm, and suppose that $J = \{f_1, \ldots, f_m\}$ is any other maximal element of $\mathcal{I}$. Without loss of generality, assume that

$$w(e_1) \leq w(e_2) \leq \cdots \leq w(e_n) \quad \text{and}$$
$$w(f_1) \leq w(f_2) \leq \cdots \leq w(f_m).$$

Assume that $|J| > |I|$; then, by the augmentation property, there is an element $e \in J$ and not in I such that $I \cup \{e\}$ is in $\mathcal{I}$, but this element would have been identified during execution of the greedy algorithm. By a similar argument, $|J| \not< |I|$ since, again by the augmentation property, we could find an element $e \in I$ so that $J \cup \{e\} \in \mathcal{I}$, thus J is not maximal.

Therefore, $|I| = |J|$ or, more specifically, $m = n$. Assume that $I_k = \{e_1, \ldots, e_k\}$ and $J_k = \{f_1, \ldots, f_k\}$ for $k = 1, \ldots, n$ (thus, I_1 and J_1 each has one element, I_2 and J_2 each has two elements, etc.) It now suffices to show that if

$$w(I_k) = \sum_{i=1}^{k} w(e_i),$$

then $w(I_k) \leq w(J_k)$ for all $k = 1, \ldots, n$. We proceed by induction. Since the greedy algorithm selects the element e with the smallest weight first, it is clear that $w(I_1) \leq w(J_1)$; thus, we have established the base case. Now, assume that the statement is true up through some arbitrary $k < n$. By definition, we know that $|J_{k+1}| > |I_k|$; therefore, by the augmentation property, there is some $e \in J_{k+1}$ with $e \notin I_k$ so that $I_k \cup \{e\}$ is an element of $\mathcal{I}$. It follows that $w(e_{k+1}) \leq w(e)$ because otherwise, the greedy algorithm would have chosen e instead of e_{k+1}. Furthermore, $w(e) \leq w(f_{k+1})$ since the elements of I and J are listed in ascending order and $e \in J_{k+1}$. Thus, $w(e_{k+1}) \leq w(e) \leq w(f_{k+1})$; therefore, we conclude that $w(I_{k+1}) \leq w(J_{k+1})$. The result follows by induction.

($\Leftarrow$) We proceed by contrapositive to prove that M is a matroid. Suppose that the augmentation property is not satisfied, and consider I and J in $\mathcal{I}$ with $|I| < |J|$ so that there is *no* element $e \in J$ with $e \notin I$ so that $I \cup \{e\}$ is in $\mathcal{I}$. Without loss of generality, assume that $|I| = |J| + 1$. Let $|I| = n$ and consider the following weight function

$$w(e) = \begin{cases} -(n+2) & \text{if } e \in I, \\ -(n+1) & \text{if } e \in J \setminus I, \\ 0 & \text{otherwise.} \end{cases}$$

After the greedy algorithm chooses all the elements of I, it cannot decrease the weight of the independent set because only elements that are not in J will be added. Thus, the total weight will be $-n(n+2) = -n^2 - 2n$. However, the set J has a weight of $-(n+1)(n+1) = -n^2 - 2n - 2$. Thus, any independent set containing J has a weight of at most $-n^2 - 2n - 2$. Thus, the greedy algorithm cannot identify a maximal independent set with minimum weight when the augmentation property is not satisfied. Thus, by contrapositive, we have shown that if the greedy algorithm identifies a maximal independent set with minimal weight, then M must be a matroid. This completes the proof. □

Theorem 5.67. *Let $G = (V, E)$ be a graph. Then, the hereditary system $M(G) = (E, \mathcal{I})$, where $\mathcal{I}$ is the collection of subsets that induce acyclic graphs, is a matroid.*

Proof. From Proposition 5.58, we know that $(E, \mathcal{I})$ is a hereditary system, we must simply show that it has the augmentation property. To see this, let I and J be two elements of $\mathcal{I}$ with $|I| < |J|$. Let H be the subgraph of G induced from the edge sets $I \cup J$. Let F be a spanning forest of this subgraph H that contains I. We know from Corollary 4.17 that H has a spanning subgraph, and we know that we can construct such a graph using the technique from the proof of Theorem 4.16.

Since J is acyclic, F has at least as many edges as J; therefore, there exists at least one edge e that is in forest F but that does not occur in the set I; furthermore, it must be an element of J (by construction of H). Since e is an edge in F, it follows that the subgraph induced by the set $I \cup \{e\}$ is acyclic; therefore, $I \cup \{e\}$ is an element of $\mathcal{I}$. Thus, $M(G)$ has the augmentation property and is a matroid. □

Corollary 5.68 (Theorem 5.38). *Let (G, w) be a weighted graph. Then, Kruskal's algorithm returns an MST when G is connected.*

□

Remark 5.69. Matroid theory is a very active and deep area of research in combinatorics and combinatorial optimization theory. A complete study of this field is well outside the scope of this book. See Ref. [47] for complete details on the subject.

5.9 Chapter Notes

This chapter is heavy with topics that would be covered in a computer science text or in a class on algorithms. As such, most of the names that appear are associated more with computer science. Robert Prim (of Prim's algorithm) was published in 1957 [48] but was actually an independent rediscovery of work originally done by Vojtěch Jarník [49]. Joseph Kruskal published his algorithm in 1956 in the *Proceedings of the American Mathematical Society* [50]. Ironically, Kruskal's algorithm was rediscovered independently in 1957 by Loberman and Weinberger [51]. Dijkstra's algorithm was originally published in 1959 [52], though conceived earlier [53]. The Floyd–Warshall algorithm is sometimes called the Roy–Floyd–Warshall algorithm, as it was discovered independently three times. Roy published first in 1959 [54] but in French. Floyd and Warshall both published independently in 1962 [55, 56] but in completely different contexts. The extensive rediscovery is an indicator of how active discrete mathematics was in the context of computer science in the middle of the 20th century.

Matroids were first developed by Hassler Whitney in 1935 [57] originally to study generalized notions of linear independence from linear algebra (which we will encounter later) but also to develop connections to graph theory (covered in this chapter). This work was independently developed by Takeo Nakasawa, but this was not known for some time [58]. Since then, the relationship between matroids (and their generalizations) and algebra, geometry, and graph theory has been thoroughly investigated. There is not enough room in this note to list all the contributors to this topic. James Oxley's text [47] is a thorough introduction to matroids, though shorter introductions with some additional results not presented in this chapter can be found in Ref. [59]. Matroids (and their generalizations) are beautiful because they connect many areas of discrete math and geometry together, but open questions can be fiendishly complex.

5.10 Exercises

Exercise 5.1

Prove Proposition 5.9. [Hint: The proof is almost identical to the proof for BFS.]

Exercise 5.2

Show that a breadth-first spanning tree returns a tree with the property that the walk from v_0 to any other vertex has the smallest length.

Exercise 5.3
Use Prim's algorithm to find an MST for the following graph.

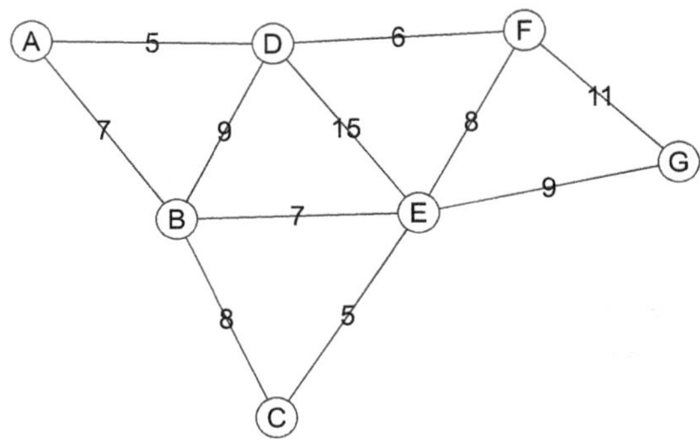

Exercise 5.4
Modify Algorithm 5 so that it returns a minimum spanning forest when G is not connected. Prove that your algorithm works.

Exercise 5.5
Use Kruskal's algorithm to determine an MST for the graph from Question 5.3.

Exercise 5.6
In the graph from Example 5.24, choose a starting vertex other than 1 and execute Prim's algorithm to show that Prim's and Kruskal's algorithms do not always add edges in the same order.

Exercise 5.7
Use Kruskal's algorithm on the following graph.

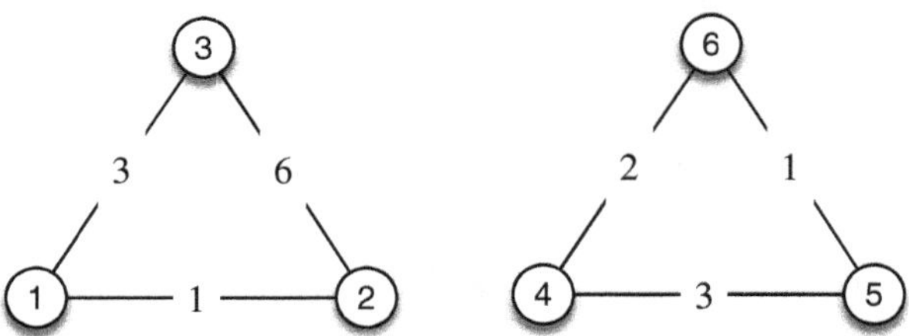

Do you obtain a minimum spanning forest?

Exercise 5.8

Experimentally compare the running time of an implementation of Kruskal's algorithm $O(|E|\log(|V|))$ to the running time of an implementation of Prim's algorithm $O(|E| + |V|\log(|V|))$. Under what circumstances might you use each algorithm? [Hint: Suppose that G has n vertices. Think about what happens when $|E|$ is big (say $n(n-1)/2$) versus when $|E|$ is small (say 0). Try plotting the two cases for various sizes of n.]

Exercise 5.9

Prove that if the edge weights of a graph are unique (i.e., no two edges share an edge weight), then that graph has only one MST.

Exercise 5.10

Use Dijkstra's algorithm to grow a Dijkstra tree for the graph in Question 5.3, starting at vertex D. Find the distance from D to each vertex in the graph.

Exercise 5.11

(Project) The A^* heuristic is a variation of Dijkstra's algorithm, which, in the worst case, defaults to Dijkstra's algorithm. It is fundamental to the study of artificial intelligence. Investigate the A^* heuristic, describe how it operates, and compare it to Dijkstra's algorithm. Create two examples for the use of the A^* heuristic, one that outperforms Dijkstra's algorithm and the other that defaults to Dijkstra's algorithm. You do not have to code the algorithms, but you can.

Exercise 5.12

(Project) Using Ref. [28] (or some other book on algorithms), implement BFS and DFS (for generating a spanning tree), Prim's, Kruskal's, and Dijkstra's algorithm in the language of your choice. Write code to generate a connected graph with an arbitrarily large number of vertices and edges. Empirically test the running time of your three algorithms to see how well the predicted running times match the actual running times as a function of the number of vertices and edges. Using these empirical results, decide whether your answer to Question 5.8 was correct or incorrect.

Exercise 5.13
The fact that the weights are not negative is never explicitly stated in the proof of the correctness of Dijkstra's algorithm, but it is used. Given this example, can you find the statement where it is critical that the weights be positive?

Exercise 5.14
Compute the running time of Steps 1–9 of Floyd's algorithm.

Exercise 5.15
Prove Proposition 5.58.

Chapter 6

An Introduction to Network Flows and Combinatorial Optimization

Remark 6.1 (Chapter goals). The goal of this chapter is to discuss flows in networks. We then apply these results to determine whether a team can advance to the playoffs in an application to sports analysis. We also discuss theoretical results on graph matching.

Remark 6.2. For the remainder of this chapter, we consider directed graphs with *no* isolated vertices and no self-loops. That is, we only consider those graphs whose incident matrices do not have any zero rows. These graphs will be connected and, furthermore, will have two special vertices v_1 and v_m, and we assume that there is at least one directed path from v_1 to v_m.

Remark 6.3. For those readers interested in the connection between flow problems and linear programming (optimization), see Chapter 12 after reading Section 6.1. An introduction to linear programming is provided in Chapter 11.

6.1 The Maximum Flow Problem

Definition 6.4 (Flow). Let $G = (V, E)$ be a digraph, and suppose $V = \{v_1, \ldots, v_m\}$ and $E = \{e_1, \ldots, e_n\}$. If $e_k = (v_i, v_j)$ is an edge, then a *flow* on e_k is a real value $x_k \geq 0$ that determines that amount of some quantity that will leave v_i and flow along e_k to v_j.

Definition 6.5 (Vertex supply and demand). Let $G = (V, E)$ be a digraph, and suppose $V = \{v_1, \ldots, v_m\}$. The *flow supply* for vertex v_i is a real value b_i assigned to v_i that quantifies that amount of flow produced at vertex v_i. If $b_i < 0$, then vertex v_i consumes flow (rather than producing it).

Definition 6.6 (Flow conservation constraint). Let $G = (V, E)$ be a digraph, and suppose $V = \{v_1, \ldots, v_m\}$ and $E = \{e_1, \ldots, e_n\}$. Let $I(i)$ be the set of edges with destination vertex v_i and $O(i)$ be the set of edges with source v_i. Then, the flow conservation constraint associated to vertex v_i is

$$\sum_{k \in O(i)} x_k - \sum_{k \in I(i)} x_k = b_i \quad \forall i. \tag{6.1}$$

Remark 6.7. Equation (6.1) states that the total flow out of vertex v_i minus the total flow into v_i must be equal to the total flow produced (or consumed) at v_i. Put more simply, excess flow is neither created nor destroyed. This is illustrated in Fig. 6.1.

Definition 6.8 (Edge capacity). Let $G = (V, E)$ be a digraph, and suppose $V = \{v_1, \ldots, v_m\}$ and $E = \{e_1, \ldots, e_n\}$. If $e_k \in E$, then its capacity is a real value $u_k \geq 0$ that determines the maximum amount of flow the edge may be assigned.

Definition 6.9 (Maximum flow problem). Let $G = (V, E)$ be a digraph, and suppose that $V = \{v_1, \ldots, v_m\}$. Let $E = \{e_1, \ldots, e_k\}$.

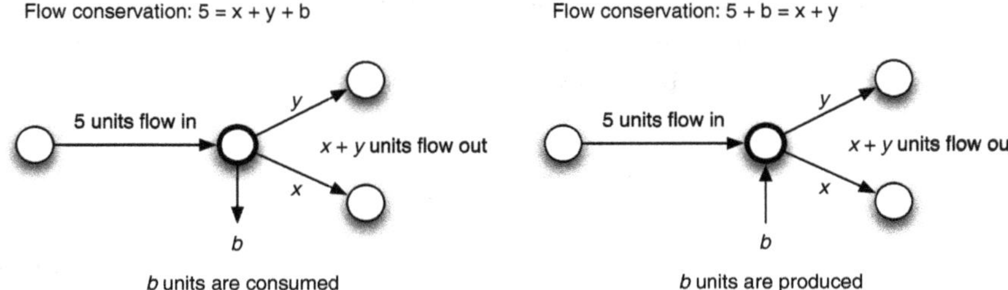

Fig. 6.1 Flow conservation is illustrated. Note that it really doesn't matter if flow is produced ($b > 0$) or consumed ($b < 0$) or neither ($b = 0$). The equations for flow conservation are identical as long as we fix a sign for b when flow is produced (in this case, $b > 0$).

Assume that the vertex supply and demand for vertices $v_2, \ldots, v_{m-1}$ is zero. That is, assume $b_2 = b_3 = \cdots = b_{m-1} = 0$. A solution to the *maximum flow problem* is the largest vertex supply b_1 so that $b_1 = -b_m$ and an edge flow $x_1, \ldots, x_k$ that satisfies all flow conservation constraints. Hence, b_1 is the largest possible amount of flow that can be forced from v_1 to v_m, and $x_1, \ldots, x_m$ is the flow that achieves this.

6.2 Cuts

Remark 6.10. Let $G = (V, E)$ be a directed graph, and suppose $V = \{v_1, \ldots, v_m\}$ and $E = \{e_1, \ldots, e_n\}$. Let V_1 be any set of vertices containing v_1 and not containing v_m, and let $V_2 = V \setminus v_1$. Immediately, we see that $v_m \in V_2$. The edges connecting vertices in V_1 with vertices in V_2 form an edge cut (see Definition 3.40); moreover, any edge cut that divides G into two components, one containing v_1 and the other containing v_m, corresponds to some sets V_1 and V_2. Thus, we refer to all such edge cuts by these generated sets; that is, (V_1, V_2) corresponds to the edge cut defined when $v_1 \in V_1$, $v_m \in V_2$, $V_1 \cap V_2 = \emptyset$, and $V_1 \cup V_2 = V$. For the remainder of this chapter, a cut refers to a cut of this type. This is illustrated in Fig. 6.2.

Definition 6.11 (Cut capacity). Let $G = (V, E)$ be a directed graph, and suppose $V = \{v_1, \ldots, v_m\}$ and $E = \{e_1, \ldots, e_n\}$. Let (V_1, V_2) be a cut separating v_1 from v_m, containing edges $e_{s_1}, \ldots, e_{s_l}$,

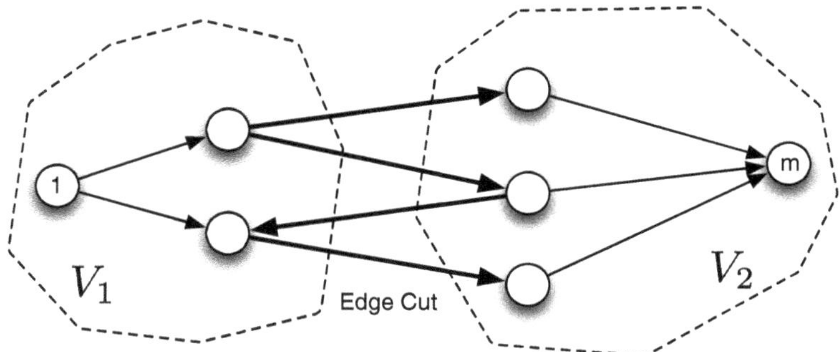

Fig. 6.2 An edge cut constructed by choosing V_1 and V_2 is illustrated. Note that $v_1 \in V_1$ and $v_m \in V_2$, separating the source from the destination.

with sources in V_1 and destinations in V_2. Here, $s_1, \ldots, s_l$ is a subset of the edge indexes $1, \ldots, n$. Then, the *capacity of the cut* (V_1, V_2) is

$$C(V_1, V_2) = \sum_{k=1}^{l} u_{s_k}. \tag{6.2}$$

That is, it is the sum of the capacities of the edges connecting a vertex in V_1 to a vertex in V_2.

Definition 6.12 (Minimum cut problem). Let $G = (V, E)$ be a directed graph, and suppose $V = \{v_1, \ldots, v_m\}$ and $E = \{e_1, \ldots, e_n\}$, with v_1 and v_m being the source and destination vertices, respectively. The *minimum cut problem* is to identify a cut (V_1, V_2) with minimum capacity $C(V_1, V_2)$.

Lemma 6.13 (Weak duality). *Let $G = (V, E)$ be a directed graph, and suppose $V = \{v_1, \ldots, v_m\}$ and $E = \{e_1, \ldots, e_n\}$. Let $(b_1^*, x_1^*, \ldots, x_k^*)$ be a solution to the maximum flow problem, where b_1^* is the maximum flow. Let (V_1^*, V_2^*) be a solution to the maximum cut problem. Then, $b_1^* \leq C(V_1^*, V_2^*)$.*

Proof. Flow conservation ensures that the total flow out of V_1^* must be equal to the total flow into V_2^* because $b_2 = b_3 = \cdots = b_{m-1} = 0$. Therefore, the maximum flow b_1^* must be forced across the edge cut (V_1^*, V_2^*). The edges leaving V_1^* and going to V_2^* have a capacity of $C(V_1^*, V_2^*)$. Therefore, it is impossible to push more than $C(V_1^*, V_2^*)$ from V_1^* to V_2^*. Consequently, $b_1^* \leq C(V_1^*, V_2^*)$. □

6.3 The Max-Flow/Min-Cut Theorem

Lemma 6.14. *In any optimal solution to a maximum flow problem, every directed path from v_1 to v_m must have at least one edge at capacity.*

Proof. By way of contradiction, suppose there is a directed path from v_1 to v_m on which the flow on each edge is not at capacity. Suppose this path contains edges from the set $J \subseteq \{1, \ldots, n\}$. That is, the flows values of this path are $x_{j_1}, \ldots x_{j_l}$, where $|J| = l$. Then,

$x_{j_i} < u_{j_i}$ for each $j_1, \ldots, j_l$. Let

$$\Delta = \min_{i \in \{1,\ldots,l\}} u_{j_i} - x_{j_i}.$$

Replacing x_{j_i} by $x_{j_i} + \Delta$ and b_1 by $b_1 + \Delta$ (and b_m by $b_m - \Delta$) maintains flow conservation and increases the maximum flow by Δ. Therefore, we could not have started with a maximum flow. This completes the proof. □

Theorem 6.15. *Let $G = (V, E)$ be a directed graph, and suppose $V = \{v_1, \ldots, v_m\}$ and $E = \{e_1, \ldots, e_n\}$. There is at least one cut (V_1, V_2) so that the flow from v_1 to v_m is equal to the capacity of the cut (V_1, V_2).*

Proof. Let $(b_1^*, \mathbf{x}^*)$ be a solution to the maximum flow problem, which we know must exist. Here, $\mathbf{x}^* = (x_1^*, \ldots, x_k^*)$.

By Lemma 6.14, we know that in this solution, every directed path from v_1 to v_m must have at least one edge at capacity. From each path from v_1 to v_m, select an edge that is at capacity in such a way that we minimize the total sum of the capacities of the chosen edges. Denote this set of edges as E'.

If E' is not yet an edge cut in the underlying graph of G, then there are some paths from v_1 to v_m in the underlying graph of G that are *not* directed paths from v_1 to v_m. In each such path, there is at least one edge directed from v_m toward v_1 (otherwise, we would have added another edge to the cut). Choose one edge from each of these paths directed from v_m to v_1 to minimize the total cardinality of edges chosen, and add these edges to E' (see Fig. 6.3).

Let V_1 be the set of vertices reachable from v_1 by a simple path in the underlying (undirected) graph $G - E'$, and let V_2 be the set of vertices reachable from v_m by a simple path in the underlying graph $G - E'$. This construction is illustrated in Fig. 6.3.

Claim 1. *Every vertex is either in V_1 or V_2 using the definition of E', and thus, the set $E' = (V_1, V_2)$ is an edge cut in the underlying graph of G.*

Proof. See Question 6.1. □

Suppose $E' = \{e_{s_1}, \ldots, e_{s_l}\}$.

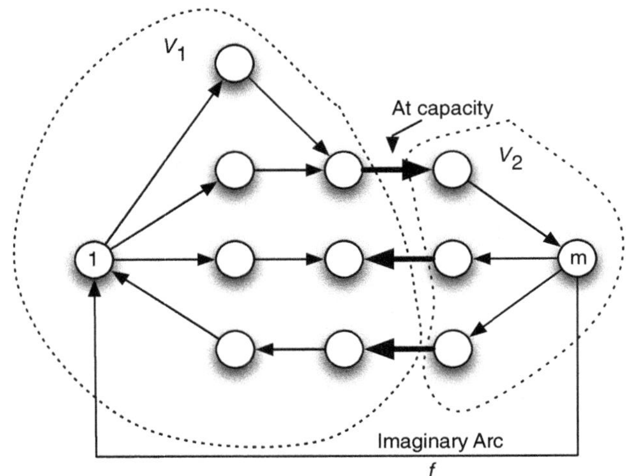

Fig. 6.3 A cut is defined as follows: In each directed path from v_1 to v_m, we choose an edge at capacity so that the collection of chosen edges has minimum capacity (and flow). If this set of edges is not an edge cut of the underlying graph, we add edges that are directed from v_m to v_1 in a simple path from v_1 to v_m in the underlying (undirected) graph of G.

Claim 2. *If there is some edge e_k with source in V_2 and destination in V_1, then $x_k = 0$.*

Proof. If $x_k \neq 0$, we could reduce this flow to zero and increase the net flow from v_1 to v_m by adding this flow to b_1. If the flow cannot reach v_1 along e_k (illustrated by the middle path in Fig. 6.3), then flow conservation ensures it must be equal to zero. □

Claim 3. *The total flow from v_1 to v_m must be equal to the capacity of the edges in E' that have source in V_1 and destination in V_2.*

Proof. We've established that if there is some edge e_k with source in V_2 and destination in V_1, then $x_k = 0$. Thus, the flow from v_1 to v_m must traverse edges leaving V_1 and entering V_2. Thus, the flow from v_1 to v_m must be equal to the capacity of the cut $E' = (V_1, V_2)$. □

Claim 3 establishes that the flow b_1^* must be equal to the capacity of a cut (V_1, V_2). This completes the proof of the theorem. □

Corollary 6.16 (Max-flow/min-cut theorem). *Let $G = (V, E)$ be a directed graph, and suppose $V = \{v_1, \ldots, v_m\}$ and $E = \{e_1, \ldots, e_n\}$. Then, the maximum flow from v_1 to v_m is equal to the capacity of the minimum cut separating v_1 from v_m.*

Proof. By Theorem 6.15, if $(b_1^*, \mathbf{x}^*)$ is a maximum flow in G from v_1 to v_m, then there is a cut (V_1, V_2) so that the capacity of this cut is equal to b_1^*. Since b_1^* is bounded above by the capacity of the minimal cut separating v_1 from v_m, the cut constructed in the proof of Theorem 6.15 *must* be a minimal capacity cut. Thus, the maximum flow from v_1 to v_m is equal to the capacity of the minimum cut separating v_1 from v_m. □

6.4 An Algorithm for Finding Optimal Flow

Remark 6.17. The proof of the max-flow/min-cut theorem we presented is a bit of a nonstandard proof technique. Most techniques are constructive; that is, they specify an algorithm for generating a maximum flow and then show that this maximum flow must be equal to the capacity of the minimal cut. In this section, we develop this algorithm and show that it generates a maximum flow and then (as a result of the max-flow/min-cut theorem) this maximum flow must be equal to the capacity of the minimum cut.

Definition 6.18 (Augment). Let $G = (V, E)$ be a directed graph, and suppose $V = \{v_1, \ldots, v_m\}$ and $E = \{e_1, \ldots, e_n\}$. Let $\mathbf{x}$ be a feasible flow in G. Consider a *simple path* $p = (v_1, e_1, \ldots, e_l, v_m)$ in the underlying graph of G from v_1 to v_m. The augment of p is the quantity

$$\min_{k \in \{1,\ldots,l\}} \begin{cases} u_k - x_k & \text{if the edge } e_k \text{ is directed toward } v_m, \\ x_k & \text{otherwise.} \end{cases} \tag{6.3}$$

Definition 6.19 (Augmenting path). Let $G = (V, E)$ be a directed graph, and suppose $V = \{v_1, \ldots, v_m\}$ and $E = \{e_1, \ldots, e_n\}$. Let $\mathbf{x}$ be a feasible flow in G. A simple path p in the underlying graph of G from v_1 to v_m is an *augmenting path* if its augment is nonzero. In this case, we say that flow $\mathbf{x}$ has an augmenting path.

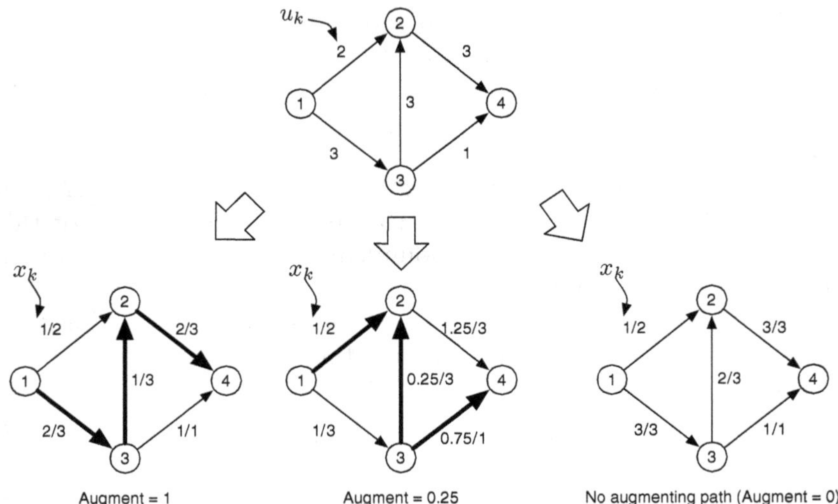

Fig. 6.4 Two flows with augmenting paths and one with no augmenting paths are illustrated.

Example 6.20. An example of augmenting paths is shown in Fig. 6.4. An augmenting path is simply an indicator that more flow can be pushed from vertex v_1 to vertex v_m. For example, in the flow on the bottom left of Fig. 6.4, we could add an additional unit of flow on the edge (v_1, v_3). This one unit could flow along edge (v_3, v_2) and then along edge (v_2, v_4). Augmenting paths that are augmenting solely because of a backward flow away from v_1 to v_m can also be used to increase the net flow from v_1 to v_m by removing flow along the backward edge.

Definition 6.21 (Path augment). If p is an augmenting path in G with augment Δ, then by *augmenting* p by Δ, we mean adding Δ to the flow in each edge directed from v_1 toward v_m and subtract Δ from the flow in each edge directed from v_m to v_1.

Example 6.22. Figure 6.5 shows the result of *augmenting* the flows shown in Example 6.20.

Remark 6.23. Algorithm 11, sometimes called the Edmonds–Karp algorithm, finds a maximum flow in a network by discovering and removing all augmenting paths.

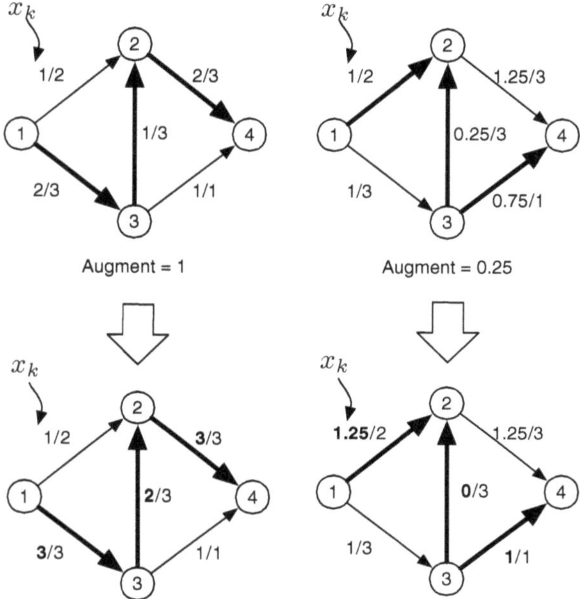

Fig. 6.5 The result of augmenting the flows shown in Fig. 6.4.

Maximum Flow Algorithm
Input: (G, u) a weighted directed graph with $G = (V, E)$, $V = \{v_1, \ldots, v_m\}$, $E = \{e_1, \ldots, e_n\}$
Initialize: $\mathbf{x} = \mathbf{0}$ {Initialize all flow variables to zero.}

(1) Find the shortest augmenting path p in G using the current flow $\mathbf{x}$
(2) **if** no augmenting path exists **then** STOP
(3) **else** augment the flow along path p to produce a new flow $\mathbf{x}$
(4) **end if**
(5) GOTO (1)

Output: $\mathbf{x}^*$

Algorithm 11: Maximum flow algorithm.

Example 6.24. We illustrate an example of the Edmonds–Karp algorithm in Fig. 6.6. Note that the capacity of the minimum cut is equal to the total flow leaving Vertex 1 and flowing to Vertex 4 at the completion of the algorithm.

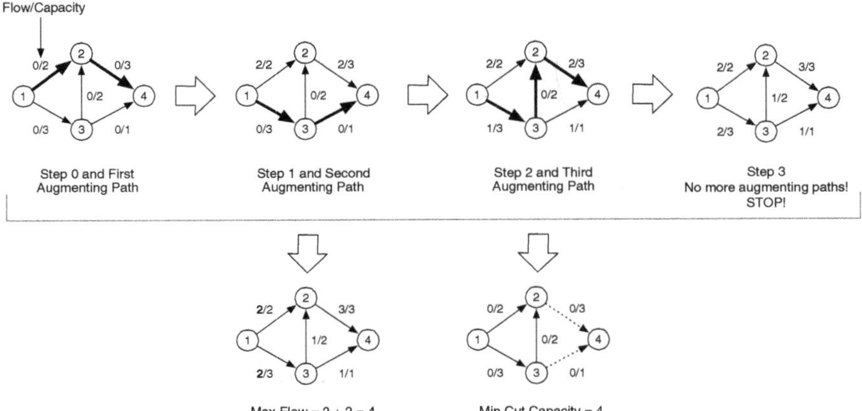

Fig. 6.6 The Edmonds–Karp algorithm iteratively augments flow on a graph until no augmenting paths can be found. An initial zero-feasible flow is used to start the algorithm. Note that the capacity of the minimum cut is equal to the total flow leaving Vertex 1 and flowing to Vertex 4.

Remark 6.25. The Edmonds–Karp algorithm is a specialization (and correction) to the Ford–Fulkerson algorithm, which does not specify how the augmenting paths in Line 1 are chosen.

Lemma 6.26. *Let $G = (V, E)$ be a directed graph, and suppose $V = \{v_1, \ldots, v_m\}$ and $E = \{e_1, \ldots, e_n\}$. Let $\mathbf{x}^*$ be a flow $\mathbf{x}^*$ that is optimal if and only if it does not have an augmenting path.*

Proof. Our proof is by abstract example. Without loss of generality, consider Fig. 6.7. Suppose there is a nonzero augment in the path shown in Fig. 6.7. If the flow f_1 is below capacity c_1, and this is the augment, then we can increase the total flow along this path by increasing the flow on each edge in the direction of v_m (from v_1) by $\Delta = c_1 - f_1$ and decreasing the flow on each edge in the direction of v_1 (from v_m) by Δ. Flow conservation at each vertex on the path is preserved since we see that

$$f_1 + f_2 - f_5 = 0 \quad \implies \quad (f_1 + \Delta) + (f_2 - \Delta) - f_5 = 0 \quad \text{and} \tag{6.4}$$

$$f_3 + f_2 - f_4 = 0 \quad \implies \quad (f_3 + \Delta) + (f_2 - \Delta) - f_4 = 0 \tag{6.5}$$

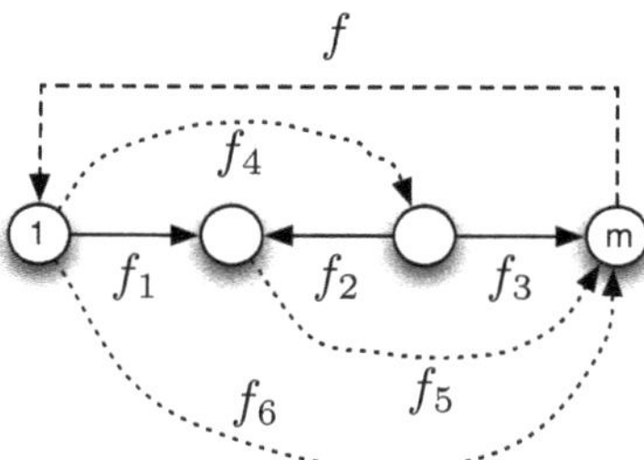

Fig. 6.7 Illustration of the impact of an augmenting path on the flow from v_1 to v_m.

as well as

$$b_1 - f_4 - f_6 - f_1 = 0 \implies (b_1 + \Delta) - f_4 - f_6 - (f_1 + \Delta) = 0 \quad \text{and} \tag{6.6}$$

$$f_5 + f_6 + f_3 + b_5 = 0 \implies f_5 + f_6 + (f_3 + \Delta) + (b_5 - \Delta) = 0. \tag{6.7}$$

Remember that $b_5 < 0$ because it is a demand.

The same argument holds if the flow on $f_2 > 0$, and this edge produces the nonzero augment. In this case, we can increase the total flow by decreasing the flow on each edge in the direction of v_1 (from v_m) by $\Delta = f_2$ and increasing the flow on each edge in the direction of v_m (from v_1) by Δ. Thus, if an augmenting path exists, the flow cannot be maximal.

Conversely, suppose we do not have a maximal flow. Then, by the max-flow/min-cut theorem, the flow across the minimal edge cut is not equal to its capacity. Thus, there is some edge in the minimal edge cut whose flow can be increased. Thus, there must be an augmenting path. This completes the proof. □

Remark 6.27. The proof of Lemma 6.26 also illustrates that Algorithm 11 maintains flow feasibility as it is executed.

Remark 6.28. A proof that Algorithm 11 terminates is a bit complicated. The main problem is showing that augmenting paths (adding or subtracting flow) does not lead to an infinite cycle of augmentations. The proof can be found in Ref. [59], where the running time is also proved.

Theorem 6.29. *Algorithm* 11 *terminates in* $O(mn^2)$ *time.* □

Theorem 6.30. *At the completion of Algorithm* 11*, there are no augmenting paths and the flow* $\mathbf{x}^*$ *is optimal.*

Proof. To see that $\mathbf{x}^*$ is feasible, note that we never increase the flow along any path by more than the maximum amount possible to ensure feasibility in all flows, and a flow is never decreased beyond zero. This is ensured in our definition of augment. We start with a feasible (zero) flow; therefore, the flow remains feasible throughout Algorithm 11.

To prove optimality, suppose at the completion of Algorithm 11, there was an augmenting path p. If we execute Line 1 of the algorithm, we will detect that augmenting path. Thus, no augmenting path exists at the conclusion of Algorithm 11, and by Lemma 6.26, $\mathbf{x}^*$ is optimal. □

Corollary 6.31 (Integral flow theorem). *If the capacities of a network are all integers, then there exists an integral maximum flow.* □

Remark 6.32. It is worth noting that the original form of Algorithm 11 did not specify which augmenting path to find. This leads to a pathological condition in which the algorithm occasionally will not terminate. This is detailed in Ford and Fulkerson's original paper and more recently in Ref. [60]. The shortest augmenting path can be found using a breadth-first search on the underlying graph. This breadth-first search is what leads to the proof of Theorem 6.29.

6.5 Applications of the Max-Flow/Min-Cut Theorem

Remark 6.33. Consider the following scenario: A baseball team wins the pennant if it obtains more wins than any other team in its division. (A similar structure can be observed in hockey, except this partially determines playoff eligibility.) At the start of the season, any team can win the pennant; however, as play continues, it occasionally becomes mathematically impossible for a team to win the pennant because of the number of losses they have incurred and the remaining schedule of games to be played. Determining whether a team can still win the pennant is an interesting mathematical problem that can be phrased as a max-flow problem. For simplicity, we ignore modern elements such as wild card spots, and we assume that if two teams tie in wins, they are still playoff (pennant) eligible and that they will play a tie-breaker game (series) in the postseason.

Example 6.34. Consider the following league standings.

				Against			
Team	**Wins**	**Losses**	**Remaining**	ATL	PHL	NY	MON
ATL	82	72	9	-	2	5	2
PHL	81	78	4	2	-	0	2
NY	79	77	5	5	0	-	0
MON	76	81	4	2	2	0	-

The against columns provide specific information on the remaining games to be played. It is clear that Montreal has been eliminated from the playoffs (or winning the division) because, with 76 games won and only four games left to play, they can never catch up to leader Atlanta. On the other hand, consider the following alternative league standings.

				Against				
Team	**Wins**	**Losses**	**Remaining**	NY	BLT	BOS	TOR	DET
NY	74	60	26	-	2	7	8	9
BLT	72	62	21	2	-	4	8	7
BOS	68	67	27	7	4	-	8	8
TOR	64	71	27	8	8	8	-	3
DET	48	87	27	9	7	8	3	-

We'd like to know if Detroit can still win the division. It certainly seems that if Detroit (amazingly) won every remaining game, it could come out ahead of New York if New York lost every game, but is that possible? It seems that the only way to figure this out is to put together all possible combinations of game wins and losses and see if there is some way Detroit can succeed in taking the pennant. This is easy for a computer (though time-consuming) and all but impossible for the average sports fan scanning her morning paper. A simpler way is to phrase the problem as a maximum flow problem. Consider the figure shown in Fig. 6.8.

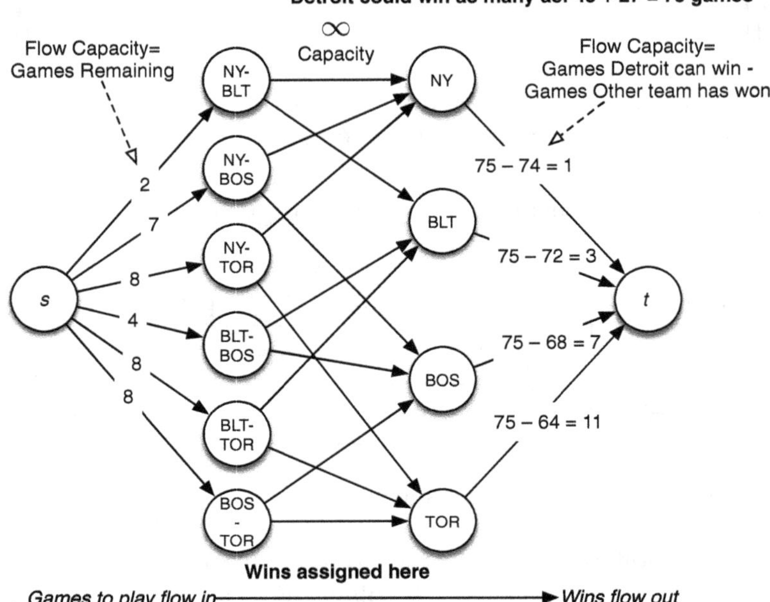

Fig. 6.8 Games to be played flow from an initial vertex s (playing the role of v_1). From here, they flow into the actual game events illustrated by vertices (e.g., NY–BOS for New York vs. Boston). Wins flow across the infinite-capacity edges to team vertices. From here, the games all flow to the final vertex t (playing the role of v_m).

In Fig. 6.8, the games to be played flow from an initial vertex s (playing the role of v_1). From here, they flow into the actual game events illustrated by vertices (e.g., NY–BOS for New York vs. Boston). Wins (and losses) occur, and the *wins* flow across the infinite-capacity edges to team vertices. From here, the games all flow to the final vertex t (playing the role of v_m). Edges going from s to the game vertices have capacity equal to the number of games left to be played between the two teams in the game vertex. This makes sense; we cannot assign more games to that edge than can be played. Edges crossing from the game vertices to the team vertices have unbounded capacity; the values we assign them will be bounded by the number of games the team plays in the game vertices anyway. Edges going from the team vertices to the final vertex t have capacity equal to the number of games Detroit can win minus the games the team whose vertex the edge leaves has already won. This tells us that for Detroit to come out on *top* (or with more wins than any other team), the number of wins assigned to a team cannot be greater than the number of wins Detroit can amass (at best). We use Detroit's possible wins because we want to know if Detroit can still win the pennant. If you were interested in another team, you would use the statistics from that team instead.

If the maximum flow in this graph fully saturates the edges leaving s, then there is an assignment of games so that Detroit can still finish first. On the other hand, if the edges connecting the team vertices to t form the minimum cut and the edges leaving s are not saturated, then there is no way to assign wins to Detroit to ensure that it wins more games than any other team (or at best ties). The maximum flow in this example is shown in Fig. 6.9. From this figure, we see that Detroit cannot make the playoffs. There is no way to assign all remaining games and for Detroit to have the most wins of any team (or to at least tie). This is evident since the edges leaving s are

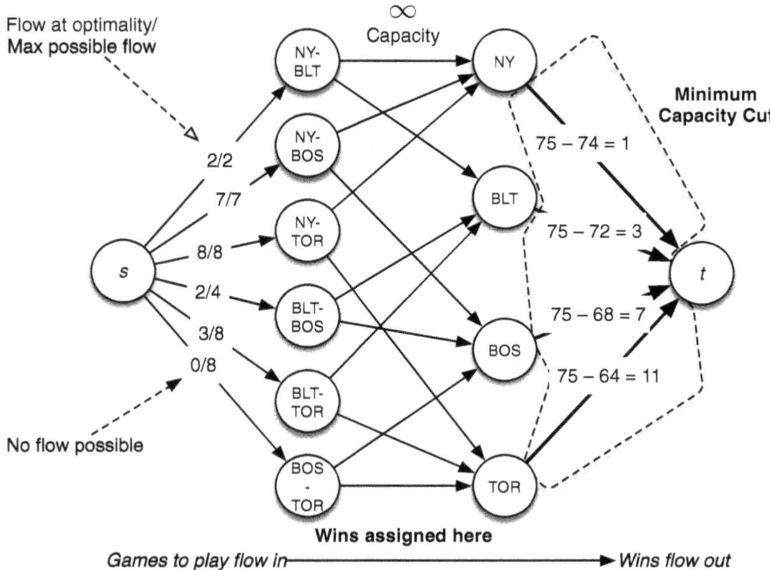

Fig. 6.9 Optimal flow was computed using the Edmonds–Karp algorithm. Note that a minimum-capacity cut consists of the edges entering t and not all edges leaving s are saturated. Detroit cannot make the playoffs.

not saturated. Note that this approach to baseball analysis was first studied by Schwartz [61].

Remark 6.35. Consider a score table for a team sport with n teams and with playoff rules like those discussed in Remark 6.33. We refer to $P(k)$ as the maximum flow problem constructed for team k ($k = 1, \ldots, n$), as in Example 6.34.

Proposition 6.36. *If the maximum flow for Problem $P(k)$ saturates all edges leaving vertex s, then team k is playoff eligible. Otherwise, team k has been eliminated.* □

6.6 More Applications of the Max-Flow/Min-Cut Theorem

Remark 6.37. The following theorem, Menger's first theorem, can be proved directly using the max-flow/min-cut theorem.

Theorem 6.38 (Menger's first theorem). *Let G be an (undirected) graph with $V = \{v_1, \ldots, v_m\}$. Then, the number of edge-disjoint paths from v_1 to v_m is equal to the size of the smallest edge cut separating v_1 from v_m.* □

Theorem 6.39 (Menger's second theorem). *Let $G = (V, E)$ be a directed graph. Let v_1 and v_2 be two nonadjacent and distinct vertices in V. The maximum number of vertex-disjoint directed paths from v_1 to v_2 is equal to the minimum number of vertices (excluding v_1 and v_2) whose deletion will destroy all directed paths from v_1 to v_2.*

Proof. We construct a new graph by replacing each vertex v in G by two vertices v' and v'' and an edge (v', v''). Each edge (v, w) is replaced by the edge (v'', w'), while each edge (u, v) is replaced by (u'', v'), as illustrated in the following.

Note that each arc of the form (v', v'') corresponds to a vertex in G. Thus, edge-disjoint paths in the constructed graph correspond to vertex-disjoint graphs in the original graph. The result follows from Menger's first theorem. □

Definition 6.40 (Matching). A matching in a graph $G = (V, E)$ is a subset M of E such that no two edges in M share a vertex in common. A matching is *maximal* if there is no other matching in G containing it. A matching has maximum cardinality if there is no other matching of G with more edges. A maximal matching is *perfect* if every vertex is adjacent to an edge in the matching.

Example 6.41. We illustrate a maximal matching and a perfect matching in Fig. 6.10.

Remark 6.42. Let $G = (V, E)$ be a graph. Recall from Definition 2.44 that a vertex cover is a set of vertices $S \subseteq V$ so that every edge in E is adjacent to at least one vertex in S.

Definition 6.43 (Minimal cover). Let $G = (V, E)$ be a graph. A vertex cover S has minimum cardinality if there is no other vertex cover S' with smaller cardinality. It is minimal if there is no vertex we can remove from S to obtain a smaller cardinality vertex S'.

Lemma 6.44. *Let $G = (V, E)$ be a graph. If M is a matching in G and C is a covering, then $|M| \leq |C|$.*

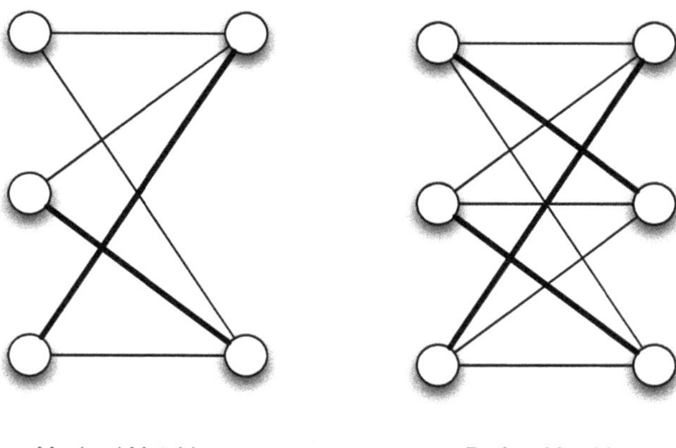

Fig. 6.10 A maximal matching and a perfect matching. Note that no other edges can be added to the maximal matching, and the graph on the left cannot have a perfect matching.

Proof. Each edge in E is adjacent to at least one element of C, meaning that C contains one end point from each edge. We may associate to each element of M either of its two end points. Let Q be the resulting set of vertices. Clearly, $|M| = |Q|$. However, since M contains only a subset of the edges in E, it is clear that Q can never have more elements than C because, at best, we can ensure that Q contains only end points of the elements of M. Thus, $|Q| \leq |C|$, which implies that $|M| \leq |C|$. Equality is achieved if M is a perfect matching. If the edges of the matching contain every vertex (e.g., it is perfect), then the covering C can be recovered by simply choosing the correct vertex from each match. □

Theorem 6.45 (König's theorem). *In a bipartite graph, the number of edges in a maximum cardinality matching is equal to the number of vertices in a minimum cardinality covering.*

Proof. Let $G = (V, E)$ be the bipartite graph with $V = V_1 \cup V_2$. Let M^* be a maximum cardinality matching for G, and let C^* be a minimum cardinality covering. First, note that $|M^*| \leq |C^*|$ by Lemma 6.44.

Construct a new graph N from G by introducing new vertices s and t so that s is adjacent to all vertices in V_1 and t is adjacent to all vertices in V_2. This is illustrated in the following.

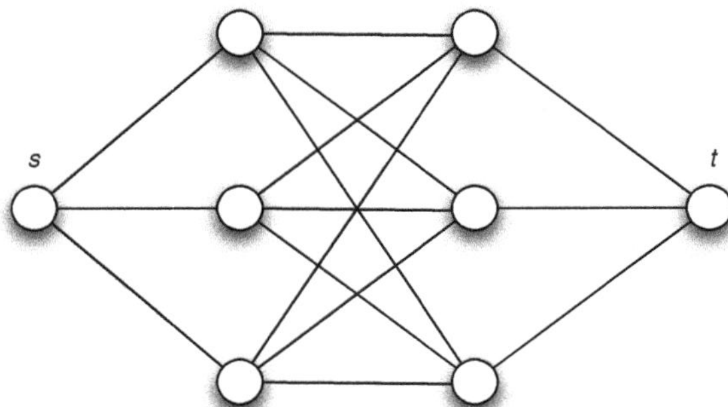

In the remainder of the proof, s will be our source (v_1) and t will be our sink (v_m). Consider a maximal (in cardinality) set P of vertex-disjoint paths from s to t. (Here, we are thinking of G as being directed from vertices in V_1 toward vertices in V_2.) Each path $p \in P$ has the form $(s, e_1, v_1, e_2, v_2, e_3, t)$, with $v_1 \in V_1$ and $v_2 \in V_2$. It is easy to see that we can construct a matching $M(P)$ from P, so for path p, we introduce the edge $e_2 = \{v_1, v_2\}$ into our matching $M(P)$. The fact that the paths in P are vertex disjoint implies that there is a one-to-one correspondence between elements in $M(P)$ and elements in P. Thus, $|P| \leq |M^*|$ since we assumed that M^* was a maximum cardinality matching.

Now, consider the smallest set $J \subset V$ whose deletion destroys all paths from s to t in N. By way of contradiction, suppose that $|J| < |C^*|$. Since we assumed that C^* was a minimal vertex cover, it follows that J is not itself a vertex cover of G, and thus, $G - J$ leaves at least one edge in G. But this edge must connect a vertex in V_1 to a vertex in V_2 because G is bipartite. Thus, $N - J$ has a path from s to t, which is a contradiction. Thus, $|C^*| \leq |J|$. Thus, we have inequalities: $|P| \leq |M^*| \leq |C^*| \leq |J|$. However, by Menger's second theorem, minimizing $|J|$ and maximizing $|P|$ implies that $|J| = |P|$, and thus, $|M^*| = |C^*|$. This completes the proof. □

Remark 6.46. König's theorem does not hold for general graphs, which can be seen by considering K_3. We discuss this more extensively in a second treatment on network flows using linear programming in Remark 12.26 in Chapter 12.

6.7 Chapter Notes

The maximum flow problem was posed by Harris and Ross as an outgrowth of the military use of operations research and optimization. In particular, Harris and Ross wanted to analyze Soviet transshipment capabilities on railways [62]. Ford and Fulkerson [63] published an incomplete solution that did not treat the possibility of an infinite loop in the augmentation process. However, their solution had all the elements of the algorithm given in this chapter, which was developed by Edmonds and Karp [64]. Interestingly, Israeli (formerly Soviet) mathematician Dinitz published a more efficient algorithm [65] two years prior to Edmonds and Karp. This algorithm, called "Dinic's" algorithm because of mispronunciation of the author's name during its popularization, has a running time of $O(|V|^2|E|)$ as compared to the running time of $O(|V||E|^2)$ of the Edmonds–Karp algorithm. However, because of the dense way Dinitz's paper had to be written to conform with Soviet journal requirements and the difficulty in East–West relations at the time, the Edmonds–Karp algorithm is the one generally taught in the West. The maximum flow problem has been and continues to be an active area of research. In particular, the notes for Chapter 12 discuss the interaction of linear programming and network flows, specifically the additional work of Orlin, who found an $O(|V||E|)$ algorithm for the maximum flow problem in 2013 [66]. In 2022, Chen *et al.* [67] posted a paper to arXiv that claims a nearly linear solution to the maximum flow problem. The running time for their algorithm is $O(|E|^{1+o(1)})$, or very close to (but not quite) $O(|E|)$. At the time of writing this chapter, that paper had not appeared in a peer-reviewed journal or conference.

6.8 Exercises

Exercise 6.1
Prove Claim 1 in the proof of Theorem 6.15.

Exercise 6.2
Prove the integral flow theorem.

Exercise 6.3
Consider the following sports standings.

Team	Wins	Loses	Remaining	vs. A	vs. B	vs. C	vs. D
A	9	5	6	-	1	3	2
B	6	10	4	1	-	2	1
C	7	6	7	3	2	-	2
D	7	8	5	2	1	2	-

Assuming that the team with the most wins will go to the playoffs at the end of the season (and ties will be broken by an extra game) and there are no wildcard spots:

(1) (1 Point) Construct a network flow problem (the picture) to determine whether Team B can still go to the playoffs.
(2) (2 Points) Determine whether Team B can still go to the playoffs.
(3) (2 Points) Determine whether Team D can still go to the playoffs.

Exercise 6.4
Prove Proposition 6.36.

Exercise 6.5

Prove Menger's first theorem. [Hint: Enumerate all edge-disjoint paths from v_1 to v_m, and replace them with directed paths from v_1 to v_m. If any edges remain undirected, then give them arbitrary direction. Assign each arc a flow capacity of 1.]

Exercise 6.6

Prove Erdös–Egerváry theorem: Let $\mathbf{A}$ be a matrix of all 0's and 1's. Then, the maximum number of 1's in matrix $\mathbf{A}$, no two of which lie in the same row or column, is equal to the minimum number of rows and columns that together contain all the 1's in $\mathbf{A}$. [Hint: Build a bipartite graph whose vertex set is composed of the row and column indices of $\mathbf{A}$. Join a row vertex to a column vertex if that position contains a 1. Now, apply a theorem from this chapter.]

Chapter 7

Coloring

Remark 7.1 (Chapter goals). The goal of this chapter is to introduce graph coloring. A large part of the chapter is devoted to proving that the problem of determining whether a graph can be colored by three colors is NP-complete. Therefore, concepts from algorithm complexity are introduced. We also prove a number of pure graph-theoretic results on graph coloring and discuss how graph coloring can be used in scheduling problems.

7.1 Vertex Coloring of Graphs

Definition 7.2 (Vertex coloring). Let $G = (V, E)$ be a graph, and let $C = \{c_1, \ldots, c_k\}$ be a finite set of *colors* (labels). A *vertex coloring* is a mapping $c : V \to C$ with the property that if $\{v_1, v_2\} \in E$, then $c(v_1) \neq c(v_2)$.

Example 7.3. We show an example of a graph coloring in Fig. 7.1. Note that no two adjacent vertices share the same color.

Definition 7.4 (k-colorable). A graph $G = (V, E)$ is k-colorable if there is a vertex coloring with k colors.

Remark 7.5. Clearly, every graph $G = (V, E)$ is $|V|$-colorable since we can assign a different color to each vertex. We are usually interested in the minimum number of colors that can be used to color a graph.

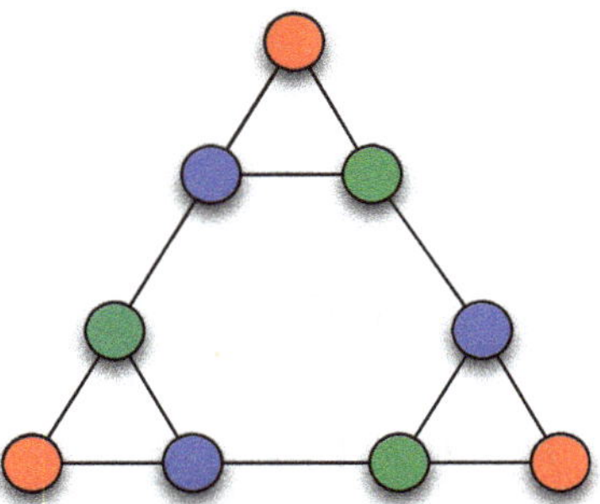

Fig. 7.1 A graph coloring. We need three colors to color this graph.

Definition 7.6 (Chromatic number). Let $G = (V, E)$ be a graph. The *chromatic number* of G, written $\chi(G)$, is the smallest positive integer k such that G is k-colorable.

Proposition 7.7. *Every bipartite graph is 2-colorable.* □

Proposition 7.8. *If $G = (V, E)$ and $|V| = n$, then*

$$\chi(G) \geq \frac{n}{\alpha(G)}, \tag{7.1}$$

where $\alpha(G)$ is the independence number of G.

Proof. Suppose $\chi(G) = k$ and consider the set of vertices $V_i = \{v \in V : c(v) = c_i\}$. Then, this set of vertices is an independent set and contains at most $\alpha(G)$ elements. Thus,

$$n = |V_1| + |V_2| + \cdots + |V_k| \leq \alpha(G) + \alpha(G) + \cdots + \alpha(G). \tag{7.2}$$

From this, we see that

$$n \leq k \cdot \alpha(G) \qquad \Longrightarrow \qquad \frac{n}{\alpha(G)} \leq k. \tag{7.3}$$

□

Proposition 7.9. *The chromatic number of K_n is n.*

Proof. From the previous proposition, we know that

$$\chi(K_n) \geq \frac{n}{\alpha(K_n)}. \tag{7.4}$$

However, $\alpha(K_n) = 1$, thus $\chi(K_n) \geq n$. From Remark 7.5, it is clear that $\chi(K_n) \leq n$. Thus, $\chi(K_n) = n$. □

Theorem 7.10. *Let $G = (V, E)$ be a graph. Then, $\chi(G) \geq \omega(G)$. That is, the chromatic number is bounded below by the size of the largest clique.* □

Theorem 7.11. *If $G = (V, E)$ is a graph with the highest degree of $\Delta(G)$, then $\chi(G) \leq \Delta(G) + 1$.*

Proof. Arrange the vertices of G in ascending order of their degree. Fix an arbitrary ordering of the colors. Assign an arbitrary color c_1 to the first vertex. Repeat this process with each vertex in order, assigning the lowest-ordered color possible. When any vertex v is to be colored, the number of colors already used cannot be any larger than its degree. At the completion of the coloring, we see that the number of colors cannot be any larger than $\Delta(G)$, thus we might require at most one extra color. Thus, $\chi(G) \leq \Delta(G) + 1$. □

Corollary 7.12. *There is at least one graph for which this bound is strict.*

Proof. Proposition 7.9 illustrates that for the complete graph, $\chi(K_n) = \Delta(K_n) + 1 = n$. □

Remark 7.13. The coloring heuristic described in Theorem 7.11 is called the greedy coloring heuristic. It is a greedy algorithm.

Proposition 7.14. *If $G = (V, E)$ is a graph and $H = (V', E')$ is a subgraph of G, then $\chi(H) \leq \chi(G)$.*

Proof. Clearly, if G is k-colorable, then so is H. Thus, $\chi(H) \leq \chi(G)$. □

Example 7.15 (Exam scheduling as a coloring problem). Suppose there are four students: Alice, Bob, Charlie, and Donna. These four students are enrolled in five different classes. Their class assignments are shown in the directed bipartite graph in Fig. 7.2 (left). For example, we see that Alice is enrolled in Classes 1, 3, and 5. The goal is to schedule final exams for these classes so that no student has a conflict; i.e., no student is required to take two or more exams simultaneously. To solve this problem, create a new graph with vertices corresponding to classes. Add an edge between two classes in this graph if and only if those two classes have at least one student in common. This is illustrated in Fig. 7.2 (right). If we imagine

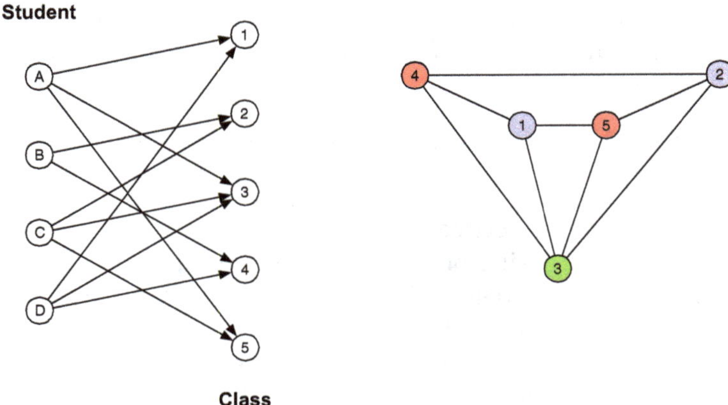

Fig. 7.2 Student class schedules can be converted into a graph with classes on the vertices and edges corresponding to classes that have at least one student in common. Scheduling exams (i.e., assigning exam slots to classes) then corresponds to coloring the graph.

that each exam slot (e.g., 9–10 AM on Thursday) is a color, then we can ask how many colors does it take to color the class graph. In this case, to create an exam schedule in which no student is required to sit two exams simultaneously, we require at most three colors or three exam slots. This is shown in Fig. 7.2 (right). This coloring can be derived using the greedy heuristic described in Theorem 7.11. The fact that we require exactly three colors follows from the fact that K_3 is a subgraph of the class graph and Proposition 7.14.

This approach to exam scheduling (or similar problems) could be generalized to much larger settings. However, as we will see in the following sections, it is generally very hard to find an optimal coloring for an arbitrary graph. Consequently, we may have to settle for a coloring that is generated by the greedy heuristic and (in some sense) "good enough."

Remark 7.16. Before proceeding, recall the following definition: The graph $K_{m,n}$ is the complete bipartite graph consisting of the vertex set $V = \{v_{11}, \ldots, v_{1m}\} \cup \{v_{21}, \ldots, v_{2n}\}$ and having an edge connecting every element of V_1 to every element of V_2. We state the following lemma without proof. A very accessible proof can be found in Ref. [68].

Lemma 7.17 (Chartrand and Kronk, 1961). Let $G = (V, E)$ be a graph. If every depth-first search tree generated from G is a Hamiltonian path, then G is either a cycle, or a complete graph, or a complete bipartite graph. □

Remark 7.18. The proof of the following theorem is based on the one from Ref. [69]. The proof is tricky and long, so it is probably best to read it slowly.

Theorem 7.19 (Brooks, 1941). If $G = (V, E)$ is connected and is neither a complete graph nor an odd cycle, then $\chi(G) \leq \Delta(G)$.

Proof. Suppose G is not regular. Choose a vertex $v_0 \in V$ with a degree of $\delta(G)$ (the smallest degree in G) and construct a depth-first search tree T starting from v_0. We now apply the following algorithm for coloring G: At step k, choose a leaf from the subtree T_k of T induced by the set of uncolored vertices. (Note that $T_0 = T$.) Color the leaf with the lowest possible index color from the set $\{c_1, \ldots, c_{\Delta(G)}\}$. In this way, the last vertex to be colored will be v_0. At each step, when $v \neq v_0$ is about to be colored, it must be adjacent to at most $\deg(v) - 1$ colored vertices. To see this, note that v is the leaf of a tree. So, *in* T, it must have a degree of 1. Thus, it is adjacent to at least one uncolored vertex that is not currently a leaf of T_k. Thus, since v is adjacent to at most $\deg(v) - 1 \leq \Delta(G) - 1$ vertices, it follows that v can be colored from $\{c_1, \ldots, c_{\Delta(G)}\}$. At last, when v_0 is colored, it is adjacent to $\delta(G) \leq \Delta(G) - 1$ colored vertices, and thus, we may choose a color from $\{c_1, \ldots, c_{\Delta(G)}\}$. Thus, G is $\Delta(G)$-colorable.

Now, suppose that G is regular. There are two possibilities: (i) G contains a cut vertex v_0; or (ii) G does not contain a cut vertex. Consider case (i) and suppose we remove v_0 to obtain several connected components. If we add v_0 back in to each of these components, then these components are not regular and each is colorable using at most $\Delta(G)$ colors by our previous result. If we arrange each of these colorings so that v_0 is colored with color c_1, then clearly the original graph is itself colorable using at most $\Delta(G)$ colors.

We are now left with case (ii) above. Consider a depth-first search tree T of G initialized from some vertex v_0 (it does not matter now which vertex is chosen since all vertices have the same degree). If T is a Hamiltonian path, then by Lemma 7.17, G is either the complete graph, a cycle, or a complete bipartite graph. By assumption, G is not a complete graph, nor is it an odd cycle. If G is an even cycle, then order the vertices from 1 to $|V|$ and color the odd-numbered vertices with c_1 and the even-numbered vertices with c_2. This is clearly a two coloring of G and $\Delta(G) = 2$. On the other hand, if G is a complete bipartite graph, then by Proposition 7.7, G is 2-colorable and $2 \leq \Delta(G)$ (because G has at least three vertices since $K_{1,1} = K_2$, which we discount by assumption).

Finally, suppose that T is not a Hamiltonian path. Then, there is some vertex $v \in T$ with a degree of at least 3. Suppose that u and w are two vertices adjacent to v that were added to T after v in the depth-first search. From this, we know that u and w are not adjacent (if they were, one would not be adjacent to v). Thus, we can color v and w with color c_1, and then, in the depth-first tree from v, we repeat the same process of coloring vertices that we used in the non-regular case. When we are about to color v, since we have used only at most $\Delta(G) - 1$ colors to color the neighbors of v (since w and u share a color), we see that there is one color remaining for v. Thus, G is $\Delta(G)$-colorable. □

7.2 Some Elementary Logic and NP-Completeness

Remark 7.20. Our goal in this section is to provide a simple definition of propositional calculus and the satisfiability problem so that we can use it to prove that determining whether a graph is 3-colorable is NP-complete. The majority of the discussion on logic is taken from Ref. [70].

Definition 7.21. The propositional connectives are: and ($\wedge$), or ($\vee$), and not $\neg$. The connectives $\wedge$ and $\vee$ are binary, while $\neg$ is a unary connective.

Definition 7.22. A propositional language L is a set of propositional atoms $x_1, x_2, x_3, \ldots$. An atomic formula consists of a propositional atom.

Example 7.23. A propositional atom might be the statement: "It is raining." (x_1) or "It is cloudy." (x_2).

Definition 7.24. An L-formula is generated inductively as follows:

(1) Any atomic formula is an L-formula.
(2) If ϕ_1 and ϕ_2 are two L-formulae, then $\phi_1 \wedge \phi_2$ is an L-formula.
(3) If ϕ_1 and ϕ_2 are two L-formulae, then $\phi_1 \vee \phi_2$ is an L-formula.
(4) If ϕ is an L-formula, then $\neg\phi_1$ is an L-formula.

Example 7.25. Continuing from Example 7.23, we might have the formula $x_1 \wedge x_2$, meaning "It is cloudy and it is raining."

Example 7.26. If x_1, x_2, and x_3 are propositional atoms, then $x_1 \wedge (\neg x_2 \vee x_3)$ is an L-formula.

Definition 7.27. An L-assignment is a mapping $M : L \to \{T, F\}$ that assigns to each propositional atom the value of `TRUE` (T) or `FALSE` (F).

Remark 7.28. The following proposition follows directly from induction on the number of connectives in an L-formula.

Proposition 7.29. *Given an L-assignment, there is a unique valuation v_M of any formula so that if ϕ is an L-formula $v_M(\phi) \in \{T, F\}$ given by the following:*

(1) *If ϕ is atomic, then $v_M(\phi) = M(\phi)$.*
(2) *If $\phi = \phi_1 \vee \phi_2$, then $v_M(\phi) = F$ if and only if $v_M(\phi_1) = F$ and $v_M(\phi_2) = F$. Otherwise, $v_M(\phi) = T$.*
(3) *If $\phi = \phi_1 \wedge \phi_2$, then $v_M(\phi) = T$ if and only if $v_M(\phi_1) = T$ and $v_M(\phi_2) = T$. Otherwise, $v_M(\phi) = F$.*
(4) *If $\phi = \neg\phi_1$, then $v_M(\phi) = T$ if and only if $v_M(\phi_1) = F$. Otherwise, $v_M(\phi) = T$.* □

Example 7.30. Consider the formula $x_1 \wedge (\neg x_2 \vee x_3)$. If $M(x_1) = F$ and $M(x_2) = M(x_3) = T$, then $v_m(\neg x_2) = F$, $v_M(\neg x_2 \vee x_3) = T$ and $v_M(x_1 \wedge (\neg x_2 \vee x_3)) = F$.

Definition 7.31 (Satisfiable). An L-formula ϕ is *satisfiable* if there is some L-assignment M so that $v_M(\phi) = T$. A set of formulae S is satisfiable if there is some L-assignment M so that for every $\phi \in S$, $v_M(\phi) = T$. That is, every formula in S evaluates to true under the assignment M.

Example 7.32. The formula $x_1 \wedge (\neg x_2 \vee x_3)$ is satisfiable when we have $M(x_1) = T$ and $M(x_2) = M(x_3) = T$.

Definition 7.33 (3-satisfiability). Suppose we consider a (finite) set of formulae S with the following properties:

(1) Every formula contains exactly three atoms or their negations.
(2) The atoms (or their negations) are connected by *or* ($\vee$) connectives.

For any arbitrary S, the question of whether S is satisfiable is called the 3-satisfiability problem or $3 - \text{SAT}$.

Example 7.34. Suppose S consists of the formulae:

(1) $x_1 \vee \neg x_2 \vee x_3$; and
(2) $x_4 \vee x_1 \vee \neg x_3$.

Then, the question of whether S is satisfiable is an instance of $3 - \text{SAT}$.

Remark 7.35. Note that we can express each $3 - \text{SAT}$ problem as a problem of satisfiability of one formula. In our previous example, we are really attempting to determine whether

$$(x_1 \vee \neg x_2 \vee x_3) \wedge (x_4 \vee x_1 \vee \neg x_3)$$

is satisfiable. This is the way $3 - \text{SAT}$ is usually expressed. A formula of this type, consisting of a collection of many "or" formulae combined with "and's" is said to be in conjunctive normal form (CNF). As a result, this is sometimes called $3 - \text{CNF} - \text{SAT}$.

Remark 7.36 (NP-completeness). A true/false question (problem) P is NP-complete if (i) the answer can be *checked* (but not necessarily obtained) in polynomial time and (ii) if there is a transformation of any other true/false question that can be checked in polynomial time into P that can be accomplished in polynomial time. More details can be found in Ref. [28].

This definition can be a little hard to understand, and in a sense, the rest of the chapter will be used to build an understanding of what we mean. The essential idea is that it is easy to check a solution to any NP-complete problem. For example, if a graph coloring is provided, it is easy to check that it correctly colors a graph and to count the number of colors it has. But it may not be easy to find that coloring.

The second part deals with the "completeness." It simply says that "if you could solve one NP-complete problem, then you can solve them all." This process of transforming one problem into another is called *reduction*, and in this context, it is only useful if the reduction can be accomplished efficiently (in polynomial time).

Remark 7.37. We state, but do not prove, the following theorem, which was shown in Karp's original 21 NP-complete problems [71].

Theorem 7.38. *The problem of deciding* $3 - SAT$ *is* NP*-complete.* □

Remark 7.39. What the previous theorem means is that (unless P = NP), any algorithm that produces an L-assignment satisfying a set of L-formulae S composed of formulae of the type from Definition 7.33 or determines that one does not exist may take a very long time to run, but the answer it gives can be verified in a polynomial number of operations in the number of atoms and the size of S, as illustrated by Proposition 7.29.

7.3 NP-Completeness of k-Coloring

Remark 7.40. Our goal in this section is to prove that the problem of determining whether a graph is 3-colorable is NP-complete. We do this by showing that there is a *polynomial time reduction* to the 3 – SAT problem. What this means is that given an instance of the 3 – SAT problem, we show that we can construct a graph that is 3-colorable if and only if the 3 – SAT instance is satisfiable in a polynomial amount of time (as a function of the size of S). Thus, if we could solve the 3-colorability problem in a polynomial amount of time, we'd be able to solve 3 – SAT in a polynomial amount of time. This contradiction implies that 3-coloring is NP-complete.

Theorem 7.41. *Deciding whether a graph is 3-colorable is* NP-*complete.*

Proof. Consider an instance of 3 – SAT with a finite set of formulae S. We construct a graph G that is 3-colorable if and only if S is satisfiable, and we argue that this construction can be completed in a number of operations that is a polynomial function of the number of formulae in S and the number of atoms in the underlying propositional language.

We initialize the graph G with three vertices $\{T, F, B\}$ that form a complete subgraph. Here, T will be a vertex representing `TRUE`, F will be the vertex representing `FALSE`, and B is a bridge vertex. Without loss of generality, assume we color T green, F red, and B blue. This is shown in Fig. 7.3.

For each propositional atom x_i in the logical language L we are considering, add two vertices v_i and v_i' to G. Add an edge $\{v_i, v_i'\}$ to G, as well as edges $\{v_i, B\}$ and $\{v_i', B\}$. This ensures that (i) v_i and v_i' cannot have the same color and (ii) neither v_i nor v_i' can have the same color as vertex B. Thus, one must be colored green and the other red. That means either x_i is true (corresponding to v_i colored green) or $\neg x_i$ is true (corresponding to v_i' colored green). This is illustrated in Fig. 7.4.

By assumption, each formula ϕ_j in S has the structure $\alpha_j(x_{j_1}) \vee \beta_j(x_{j_2}) \vee \gamma_j(x_{j_3})$, where $\alpha_j(x_{j_1}) = x_{j_i}$, if $\phi_j = x_{j_1} \vee \cdots$ and $\alpha_j(x_{j_1}) = \neg x_{j_1}$ if $\phi_j = \neg x_{j_1} \vee \cdots$. The effects of β_j and γ_j are defined similarly. Add the five vertices $t_{j_1}, t_{j_2}, \ldots, t_{j_5}$ to the graph with the properties that

(1) t_{j_1}, t_{j_2}, and t_{j_4} form the subgraph K_3;
(2) t_{j_4} is adjacent to t_{j_5};
(3) t_{j_5} is adjacent to t_{j_3};

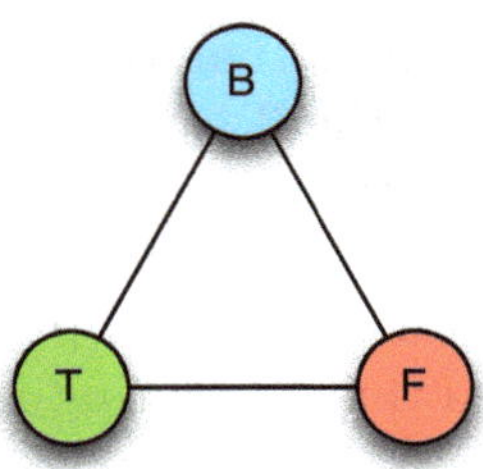

Fig. 7.3 At the first step of constructing G , we add three vertices $\{T, F, B\}$ that form a complete subgraph.

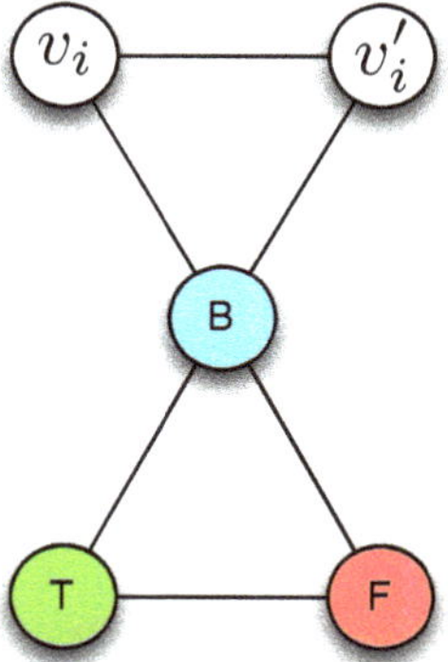

Fig. 7.4 At the second step of constructing G, we add two vertices v_i and v_i' to G and an edge $\{v_i, v_i'\}$.

(4) both t_{j_5} and t_{j_3} are adjacent to T, and

(a) if $\alpha_j(x_{j_1}) = x_{j_1}$, then t_{j_1} is adjacent to v_{j_1}, otherwise t_{j_1} is adjacent to v'_{j_1},

(b) if $\beta_j(x_{j_2}) = x_{j_2}$, then t_{j_2} is adjacent to v_{j_2}, otherwise t_{j_2} is adjacent to v'_{j_2},

(c) if $\gamma_j(x_{j_3}) = x_{j_3}$, then t_{j_3} is adjacent to v_{j_3}, otherwise t_{j_3} is adjacent to v'_{j_3}.

This construction is illustrated in Fig. 7.5 for the case when $\phi_j = x_{j_1} \vee x_{j_2} \vee x_{j_3}$. We must now show that there is a 3-coloring for this graph just in case S is satisfiable. Without loss of generality, we show the construction for the case when $\phi_j = x_{j_1} \vee x_{j_2} \vee x_{j_3}$. All other cases follow by an identical argument with a modified graph structure. For the remainder of this proof, let ν be a valuation function.

Claim 1. *If $\nu(x_{j_1}) = \nu(x_{j_2}) = \nu(x_{j_3}) =$ **FALSE**, then G is not 3-colorable.*

Proof. To see this, observe that either t_{j_1} or t_{j_2} must be colored blue and the other green since v_1, v_2, and v_3 are colored red. Thus, t_{j_4} must be colored red. Furthermore, since v_{j_3} is colored red, it follows that t_{j_3} must be colored blue. But then, t_{j_5} is adjacent to a green vertex (T), a red vertex (t_{j_4}), and a blue vertex t_{j_3}. Thus, we require a fourth color. This is illustrated in Fig. 7.6(a). □

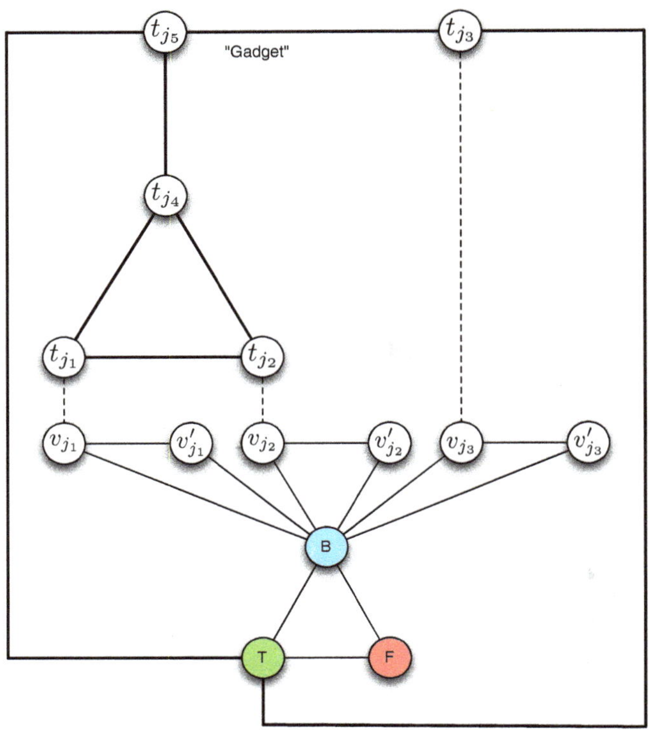

Fig. 7.5 At the third step of constructing G, we add a "gadget" that is built specifically for term ϕ_j.

Claim 2. *If $\nu(x_{j_1}) =$ TRUE or $\nu(x_{j_2}) =$ TRUE or $\nu(x_{j_3}) =$ TRUE, then G is 3-colorable.*

Proof. The proof of the claim is illustrated in Fig. 7.6(b)–(h). □

Our two claims show that by our construction of G that G is 3-colorable if and only if every every formula of S can be satisfied by some assignment of `TRUE` or `FALSE` to the atomic propositions. (It should be clear that variations of Claims 1 and 2 are true by symmetry arguments for any other possible value of ϕ_j; e.g. $\phi_j = x_{j_1} \vee \neg x_{j_2} \vee x_{j_3}$.) If we have n formulae in S and m atomic propositions, then G has $5n + 2m + 3$ vertices and $3m + 10n + 3$ edges, and thus, G can be constructed in a polynomial amount of time from S. It follows at once that since $3 - SAT$ is NP-complete,

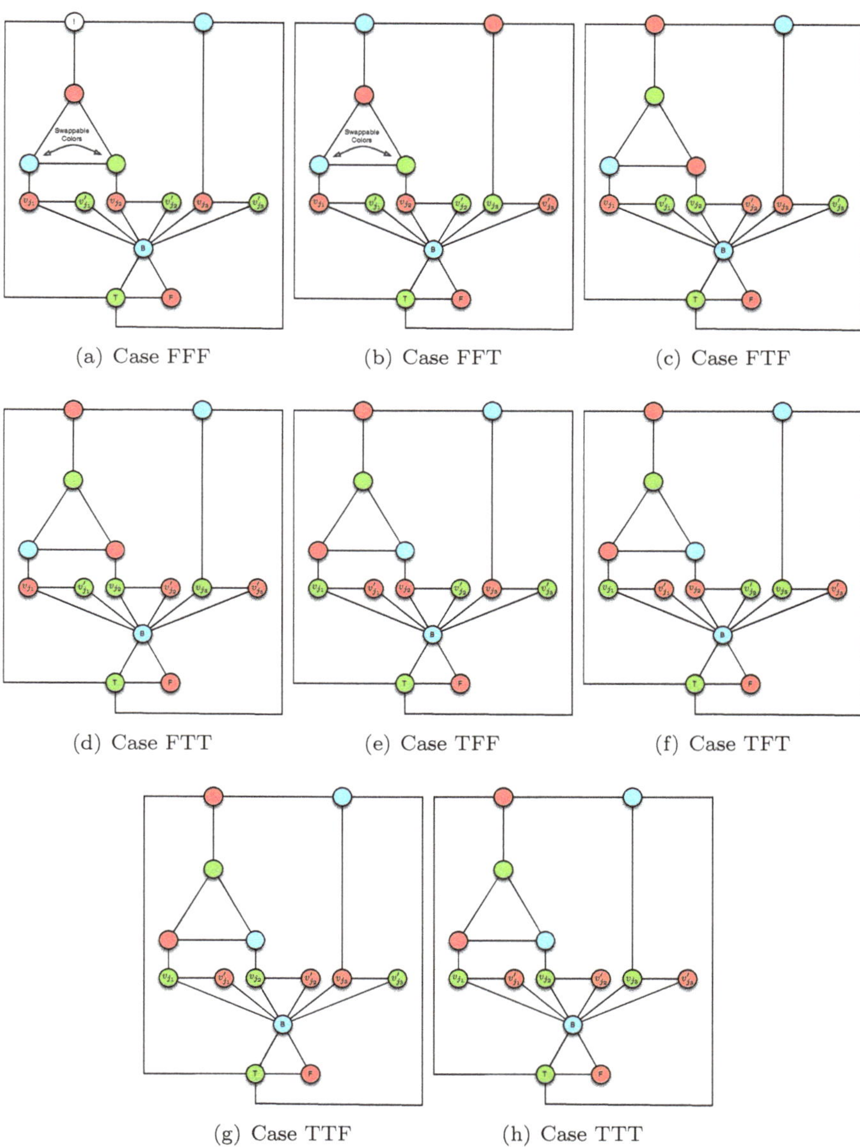

(a) Case FFF (b) Case FFT (c) Case FTF

(d) Case FTT (e) Case TFF (f) Case TFT

(g) Case TTF (h) Case TTT

Fig. 7.6 When ϕ_j evaluates to false, the graph G is not 3-colorable, as illustrated in Subfigure (a). When ϕ_j evaluates to true, the resulting graph is colorable. By the label TFT, we mean $v(x_{j_1}) = v(x_{j_3}) = \texttt{TRUE}$ and $v_{j_2} = \texttt{FALSE}$.

so is the question of whether an arbitrary graph is 3-colorable. This completes the proof. □

Corollary 7.42. *For an arbitrary k, deciding whether a graph is k-colorable is* NP*-complete.* □

7.4 Graph Sizes and k-Colorability

Remark 7.43. It is clear from Theorem 7.10 that graphs with arbitrarily high chromatic numbers exist. What is interesting is that we can induce such graphs without the need to induce cliques. In particular, we can show constructively that graphs with arbitrarily large girths exist, and these graphs have large chromatic numbers.

Lemma 7.44. *In a k-coloring of a graph $G = (V, E)$ with $\chi(G) = k$, there is a vertex of each color that is adjacent to vertices of every other color.*

Proof. Consider any k-coloring of the graph in question. For any color c_i ($i \in \{1, \ldots, k\}$), there is at least one vertex v with color c_i whose color cannot be changed. (If not, then we would repeat the process of recoloring vertices colored c_i until we need only $k - 1$ colors.) Now, suppose this v is not adjacent to $k - 1$ vertices of a color other than c_i. Then, we could recolor v with a different color, contradicting our assumption on v. Thus, we see v must have a degree of $k - 1$ and is adjacent to a vertex colored with each color in $\{c_1, \ldots, c_k\}$ other than c_i. □

Theorem 7.45. *For any positive k, there exists a triangle-free graph with a chromatic number of k.*

Proof. For $k = 1$ and $k = 2$, clearly K_1 and K_2 satisfy the criteria of the theorem. We proceed by induction. Assuming that the statement is true up to some arbitrary k, we show that the result is true for $k + 1$. Let G_k be a *triangle-free* graph with $\chi(G_k) = k$. (That is, G_k does not contain a clique since it does not contain a subgraph isomorphic to K_3.) Suppose that $V_k = \{v_1, \ldots, v_n\}$ are the vertices of G_k. We construct G_{k+1} in the following way:

(1) Add $n + 1$ vertices to G_k u_1, corresponding to v_1, through u_n, corresponding to v_n, and an extra vertex v.

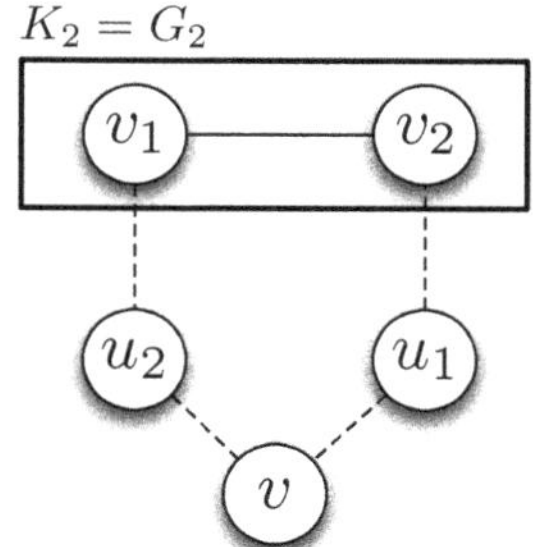

Fig. 7.7 Constructing G_3 from G_2.

(2) Add an edge from each u_i to v to form a star graph (as a subgraph of G_{k+1}.
(3) Add an edge from u_i to each *neighbor* of v_i. (That is, u_1 becomes adjacent to $v_2, \ldots, v_n$.)

This is illustrated for constructing G_3 from G_2 in Fig. 7.7.

Claim 3. *The graph G_{k+1} contains no triangles.*

Proof. The set $U = \{u_1, \ldots, u_n\}$ is an independent set. Thus, any subgraph of G_{k+1} isomorphic to K_3 must contain *at most* one element of U. Therefore, suppose that the vertices u_i, v_j, and v_k form a triangle; i.e., there are edges $\{u_i, v_j\}$, $\{u_i, v_k\}$, and $\{v_j, v_k\}$. Then, since u_i is adjacent to v_j and v_k in G_{k+1}, it follows that v_j is a neighbor of v_i and v_k is a neighbor of v_i; therefore, the edges $\{v_i, v_j\}$, $\{v_i, v_k\}$, and $\{v_j, v_k\}$ exist, and thus, there is a triangle in G_k, contradicting our inductive hypothesis. □

It now suffices to show that $\chi(G_{k+1}) = k + 1$. It is clear that G_{k+1} is at least $k+1$-colorable since any k-coloring of G_k can be extended to G_{k+1} by coloring each u_i with the same color as v_i and then coloring v with a $k+1^{\text{st}}$ color. Now, suppose that G_{k+1} is k-colorable. Applying Lemma 7.44, there is a vertex v_i having color c_1 that is adjacent to vertices having every other color. Since u_i has the same neighbors as v_i, it follows that u_i *must* also be colored c_1. Thus, all k colors appear in the vertices of u_i. But, since each vertex u_i is adjacent to v, there is no color available for v, and thus, G_{k+1} must have a chromatic number of $k + 1$. □

7.5 Chapter Notes

Graph-coloring problems originate in the coloring of planar graphs, which are not discussed in this book (in order to make more room for algebraic graph theory). A graph is *planar* if it can be drawn on a sheet of paper so that none of its edges cross. Edges do not need to be drawn as straight lines but may curve to avoid intersecting other edges. It is an interesting exercise to show that K_4 is planar whereas K_5 is not. A simple proof relies on the Euler polyhedral formula relating the number of edges, faces, and vertices of a polyhedron. Ref. [24] has a well-written introduction to planar graphs.

Planar graphs were studied in relation to the *four color theorem*: Any geopolitical map can be colored with at most four colors in a way that no adjacent states share a color. This problem was first introduced by Cayley to the London Mathematical Society [72]. The British mathematician Alfred Kempe wrote a false proof, as shown by fellow mathematician Percy Heawood, who constructed a proof of the five color theorem in the same paper. The problem remained open until 1976, when the four color theorem was proved by Appel and Haken (see Ref. [73] for the announcement and discussion by the authors). Interestingly, this was an early (perhaps the earliest) computer-aided proof in mathematics, which set the stage for later computer-assisted proof methods [74].

There has been a substantial amount of crossover between graph coloring and abstract algebra with the introduction of the chromatic polynomial by Birkhoff [75], which he used to study planar graph coloring. The work and its extensions are summarized in Ref. [76], which also discusses the generalizations of Whitney [77]. The chromatic polynomial $P_G(k)$ can be used to count the number of proper k-colorings of a graph G and is generalized by the Tutte polynomial. (See Ref. [78] for a survey of graph polynomials by Tutte.) The chromatic polynomial and other graph polynomials are a major component of algebraic graph theory, which we discuss in the following chapters.

In addition to vertex colorings, one can also study the edge colorings of graphs. Gross and Yellen [19] and Diestel [24] have authored chapters on graph colorings that discuss edge coloring. Diestel [24] also proves the five color theorem.

7.6 Exercises

Exercise 7.1
Prove Proposition 7.7.

Exercise 7.2
Prove Theorem 7.10.

Exercise 7.3
Use the greedy coloring heuristic described in the proof of Theorem 7.11 to find a coloring for the Petersen graph.

Exercise 7.4
Use the algorithm described in the proof of Theorem 7.19 to compute a coloring for the Petersen graph.

Exercise 7.5
What is the chromatic number of the following graph?

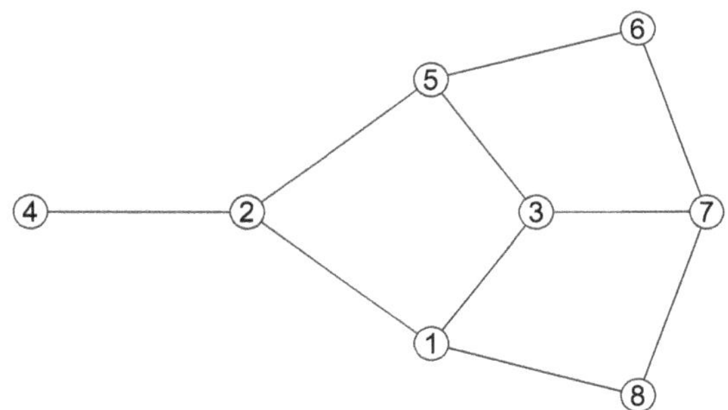

Exercise 7.6
Verify whether the L-assignment $M(x_1) = T$ and $M(x_2) = M(x_3) = T$ satisfies $x_1 \wedge (\neg x_2 \vee x_3)$.

Exercise 7.7
Construct an example of a set of L-formulae (on any set of atoms that you like) that is *not* satisfiable.

Exercise 7.8
Determine whether S in Example 7.34 is satisfiable. Illustrate with a specific L-assignment.

Exercise 7.9

A degree-k-constrained spanning tree of a graph G is a spanning tree T of G in which each vertex in T has a degree of *at most* k. Prove that the question: "Does a graph have a degree-k-constrained minimum spanning tree?" is NP-complete. To do this, you may assume the question "Does a graph have a Hamiltonian path?" is NP-complete. [Hint: Think about the case when $k = 2$.]

Exercise 7.10

Let G be a graph. Show that if G is k-colorable and we construct G' from G by adding one edge, then G' is at most $k + 1$-colorable. Show an example where this bound is strict.

Part 3
Some Algebraic Graph Theory

Chapter 8

Algebraic Graph Theory with Abstract Algebra

Remark 8.1 (Chapter goals). In this chapter, we explore graph properties using concepts from abstract algebra. We begin by discussing graph isomorphism. We then introduce basic concepts from group theory, including permutation groups. We then use these concepts to discuss graph automorphisms. This forms the first component of our exploration of algebraic graph theory.

8.1 Isomorphism and Automorphism

Definition 8.2 (Injective mapping). Let S and T be sets. A function $f : S \to T$ is *injective* (sometimes *one-to-one*) if for all $s_1, s_2 \in S$: $f(s_1) = f(s_2) \iff s_1 = s_2$.

Definition 8.3 (Surjective mapping). Let S and T be sets. A function $f : S \to T$ is *surjective* (sometimes *onto*) if for all $t \in T$, there exists an $s \in S$ such that $f(s) = t$.

Definition 8.4 (Bijective mapping). Let *bijective* if f is both injective and surjective.

Remark 8.5. An injection, a surjection, and a bijection are illustrated in Fig. 8.1.

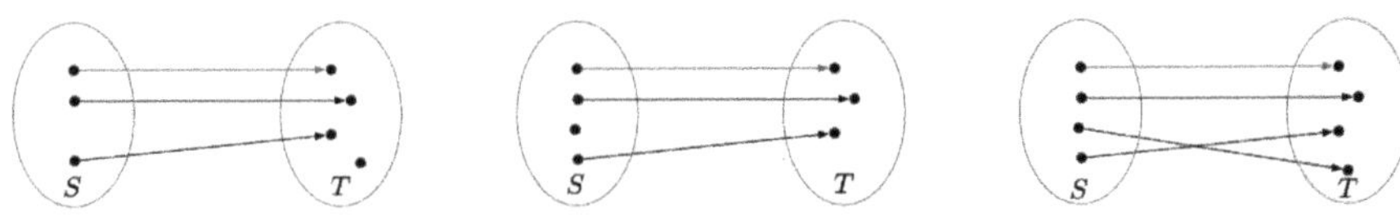

Fig. 8.1 An injection (left), a surjection (middle), and a bijection (right) are illustrated visually.

8.2 Graph Isomorphism

Definition 8.6 (Graph isomorphism). Let $G = (V, E)$ and $G' = (V', E')$. The graphs G and G' are *isomorphic* if there is a bijective mapping $f : V \to V'$ such that for all $v_1, v_2 \in V$, we have

$$\{v_1, v_2\} \in E \qquad \iff \qquad \{f(v_1), f(v_2)\} \in E'. \tag{8.1}$$

In this case, the mapping f is called a *graph isomorphism.* If G and G' are isomorphic, we write $G \cong G'$.

Definition 8.7. Let $G = (V, E)$ be a graph. Then, the set $\{H : H \cong G\}$ is called the *isomorphism type* (or *isomorphism class*) of G.

Theorem 8.8 (Graph invariant theorem). *Suppose $G = (V, E)$ and $G' = (V', E')$ are graphs, with $G \cong G'$ and $f : V \to V'$ the graph isomorphism between the graphs. Further suppose that the degree sequence of G is $\mathbf{d}$ and the degree sequence of G' is $\mathbf{d}'$. Then:*

(1) $|V| = |V'|$ *and* $|E| = |E'|$;
(2) *for all* $v \in V$, $\deg(v) = \deg(f(v))$;
(3) $\mathbf{d} = \mathbf{d}'$;
(4) *for all* $v \in V$, $\mathrm{ecc}(v) = \mathrm{ecc}(f(v))$;
(5) $\omega(G) = \omega(G')$ *(recall that* $\omega(G)$ *is the clique number of* G*)*;
(6) $\alpha(G) = \alpha(G')$ *(recall that* $\alpha(G)$ *is the independence number of* G*)*;
(7) $c(G) = c(G')$ *(recall that* $c(G)$ *is the number of components of* G*)*;
(8) $\mathrm{diam}(G) = \mathrm{diam}(G')$;
(9) $\mathrm{rad}(G) = \mathrm{rad}(G')$;
(10) *the girth of* G *is equal to the girth of* G';
(11) *the circumference of* G *is equal to the circumference of* G'.

Remark 8.9. The proof of Theorem 8.8 is long and should be clear from the definition of isomorphism. Isomorphism is really just a way of *renaming* vertices; we assume that the vertices in the graph G are named from the set V, while the vertices in the graph G' are named from the set V'. If the graphs are identical except for the names we give the vertices (and thus the names that appear in the edges), then the graphs are isomorphic, and all structural properties are preserved as a result of this.

Remark 8.10. The inverse of Theorem 8.8 does not hold. We illustrate this in Example 8.11.

Example 8.11. Given two graphs G and G', we can see by example that the degree sequence does not uniquely specify the graph G, and thus, if G and G' have degree sequences $\mathbf{d}$ and $\mathbf{d}'$, respectively, it is necessary that $\mathbf{d} = \mathbf{d}'$ when $G \cong G'$ but not sufficient to establish isomorphism. To see this, consider the graphs shown in Fig. 8.2. It's clear that $\mathbf{d} = (2, 2, 2, 2, 2, 2) = \mathbf{d}'$, but these graphs cannot be isomorphic since they have different numbers of components.

The same is true with the other graph properties. The equality between a property of G and that same property for G' is a necessary but insufficient criterion for the isomorphism of G and G'. We will not encounter any property of a graph that provides such a necessary and sufficient condition (see Remark 8.15).

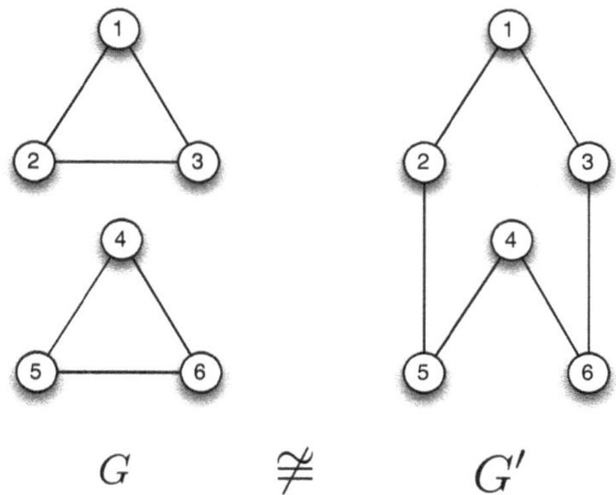

Fig. 8.2 Two graphs that have identical degree sequences but are not isomorphic.

Theorem 8.12. *Suppose that $G = (V, E)$ and $G' = (V', E')$ are graphs with $G \cong G'$ and that $f : V \to V'$ is the graph isomorphism between the graphs. If H is a subgraph of G, then $H' = f(H)$ is a subgraph of G'. (Here, $f(H)$ is the image of the subgraph H under the isomorphism f.)* □

Definition 8.13 (Graph isomorphism problem). Given two graphs $G = (V, E)$ and $G' = (V', E')$, the *graph isomorphism problem* is to determine whether or not G and G' are isomorphic.

Definition 8.14 (Subgraph isomorphism). Given two graphs $G = (V, E)$ and $H = (V', E')$, the *subgraph isomorphism problem* is to determine whether G contains a subgraph that is isomorphic to H.

Remark 8.15. In general, the subgraph isomorphism problem is NP-complete. The graph isomorphism problem (interestingly enough) is a bit of an enigma. We do not know exactly how hard this problem is to solve. We do know that it is not quite as hard as the subgraph isomorphism problem. It is worthwhile noting, however, that there is a linear time algorithm for determining the isomorphism of two trees (see p. 84 of Ref. [79]).

Definition 8.16 (Automorphism). Let $G = (V, E)$ be a graph. An *automorphism* is an isomorphism from G to itself. That is, a bijection $f : V \to V$ so that for all $v_1, v_2 \in V$, $\{v_1, v_2\} \in E \iff \{f(v_1), f(v_2)\} \in E$.

Remark 8.17 (Inverse automorphism). Recall that an isomorphism (and hence an automorphism) is a bijective function, and hence, it has a well-defined inverse. That is, if $G = (V, E)$ is a graph and $f : V \to V$ is an automorphism, then if $f(v_1) = f(v_2)$, we know that $v_1 = v_2$ (because f is injective). Furthermore, we know that for every $v_2 \in V$, there is a (unique) $v_1 \in V$ so that $f(v_1) = v_2$ (because f is surjective). Thus, if $v_2 \in V$, we can define $f^{-1}(v_2)$ to be the unique v_1 so that $f(v_1) = v_2$.

Lemma 8.18. *Let $G = (V, E)$ be a graph. Suppose that $f : V \to V$ is an automorphism. Then, $f^{-1} : V \to V$ is also an automorphism.*

Proof. The fact that f is a bijection implies that f^{-1} is itself a bijection. We know for all v_1 and v_2 in V that

$$\{v_1, v_2\} \in E \qquad \iff \qquad \{f(v_1), f(v_2)\} \in E.$$

For every vertex pair u_1 and u_2 in V, there are unique vertices v_1 and v_2 in V so that $u_1 = f(v_1)$ and $u_2 = f(v_2)$. Furthermore, by the previous observation,

$$\{u_1, u_2\} \in E \qquad \iff \qquad \{v_1, v_2\} \in E.$$

However, this means that for all u_1 and u_2 in V, we have

$$\{f^{-1}(u_1), f^{-1}(u_2)\} \in E \qquad \iff \qquad \{u_1, u_2\} \in E. \tag{8.2}$$

Thus, f^{-1} is a bijection that preserves the edge relation. This completes the proof. □

Remark 8.19. We use the next lemma in the following section. Let $f : V \to V$ and $g : V \to V$ be automorphisms of a graph G. Then, $f \circ g : V \to V$ is the function with the property that $(f \circ g)(v) = f(g(v))$. This is just function composition.

Lemma 8.20 (Composition). Let $G = (V, E)$ be a graph. Suppose that $f : V \to V$ and $g : V \to V$ are automorphisms. Then, $f \circ g$ is also an automorphism. □

8.3 Groups

Definition 8.21 (Group). A *group* is a pair $(S, \circ)$, where S is a set and $\circ : S \times S \to S$ is a binary operation so that:

(1) The binary operation $\circ$ is associative; that is, if s_1, s_2, and s_3 are in S, then $(s_1 \circ s_2) \circ s_3 = s_1 \circ (s_2 \circ s_3)$.
(2) There is a unique identity element $e \in S$ so that for all $s \in S$, $e \circ s = s \circ e = s$.
(3) For every element $s \in S$, there is an inverse element $s^{-1} \in S$ so that $s \circ s^{-1} = s^{-1} \circ s = e$.

If $\circ$ is commutative, i.e., for all $s_1, s_2 \in S$, we have $s_1 \circ s_2 = s_2 \circ s_1$, then $(S, \circ)$ is called a *commutative group* (or *abelian group*).

Definition 8.22 (Subgroup). Let $(S, \circ)$ be a group. A *subgroup* of $(S, \circ)$ is a group $(T, \circ)$ so that $T \subseteq S$. The subgroup $(T, \circ)$ shares the identity of the group $(S, \circ)$.

Example 8.23. The integers under addition $(\mathbb{Z}, +)$ is a group. We can see this because the sum of two integers is an integer. The additive identity is the number 0 because $0+n = n+0 = n$. The additive inverse of a number $n \in \mathbb{Z}$ is $-n \in \mathbb{Z}$. Addition is associative. Therefore, $(\mathbb{Z}, +)$ is a group.

Example 8.24. Consider the group $(\mathbb{Z}, +)$. If $2\mathbb{Z}$ is the set of even integers, then $(2\mathbb{Z}, +)$ is a subgroup of $(\mathbb{Z}, +)$ because the even integers are closed under addition.

Remark 8.25. Recall from Lemma 8.20 that the set of automorphisms of a graph G is closed under function composition $\circ$.

Theorem 8.26. *Let $G = (V, E)$ be a graph. Let $\text{Aut}(G)$ be the set of all automorphisms on G. Then, $(\text{Aut}(G), \circ)$ is a group.*

Proof. By Lemma 8.20, we can see that functional composition is a binary operation $\circ : \text{Aut}(G) \to \text{Aut}(G)$. Associativity is a property of functional composition since if $f : V \to V$, $g : V \to V$, and $h : V \to V$, it is easy to see that for all $v \in V$

$$\begin{aligned}((f \circ g) \circ h)(v) &= (f \circ g)(h(v)) = f(g(h(v))) = f \circ (g(h(v))) \\ &= (f \circ (g \circ h))(v). \end{aligned} \tag{8.3}$$

The identity function $e : V \to V$ defined by $e(v) = v$ for all $v \in V$ is an automorphism of V. Finally, by Lemma 8.18, each element of $\text{Aut}(G)$ has an inverse. This completes the proof. □

8.4 Permutation Groups and Graph Automorphisms

Definition 8.27 (Permutation/Permutation group). A *permutation* on a set $V = \{1, \ldots, n\}$ of n elements is a bijective mapping f from V to itself. A *permutation group* on a set V is a set of permutations with the binary operation of functional composition.

Remark 8.28. A graph automorphism is just a permutation of the vertices. Consequently, when we are studying the automorphism group of a graph, we are just studying a permutation group.

Example 8.29. Consider the set $V = \{1, 2, 3, 4\}$. A permutation on this set that maps 1 to 2, 2 to 3, and 3 to 1 can be written as $(1, 2, 3)(4)$, indicating the cyclic behavior that $1 \to 2 \to 3 \to 1$ with 4 fixed. In general, we write $(1, 2, 3)$ instead of $(1, 2, 3)(4)$ and suppress any elements that do not move under the permutation.

Example 8.30. Consider the set $V = \{1, 2, 3\}$. The symmetric group on V is the set S_3, and it contains the permutations:

(1) $(1)(2)(3) = e$ (the identity),
(2) $(12)(3) = (12)$,
(3) $(13)(2) = (13)$,
(4) $(23)(1) = (23)$,
(5) (123), and
(6) (132).

Example 8.31. For the permutation (f), taking 1 to 3, 3 to 1, 2 to 4, and 4 to 2, we write $f = (1, 3)(2, 4)$ and say that this is the *product* of $(1, 3)$ and $(2, 4)$. When determining the action of a permutation on a number, we read the permutation from right to left. Thus, if we want to determine the action of f on 2 (i.e., to compute $f(2)$), we read from right to left and see that 2 goes to 4. By contrast, if we had the permutation $g = (1, 3)(1, 2)$, then to compute $g(2)$, we see that we take 2 to 1 first and then 1 to 3; thus, 2 would be mapped to 3. We would first map the number 1 to 2 and then stop. The number 3 would be mapped to 1. Thus, we can see that $(1, 3)(1, 2)$ has the same action as the permutation $(1, 2, 3)$.

Definition 8.32 (Transposition). A permutation of the form (a_1, a_2) is called a transposition.

Definition 8.33 (Symmetric group). Consider a set V with n elements in it. The permutation group S_n contains every possible permutation of the set with n elements.

Proposition 8.34. *For each n, $|S_n| = n!$* □

Theorem 8.35. *Every permutation can be expressed as a product of transpositions.*

Proof. Consider the permutation $(a_1, a_2, \ldots, a_n)$. We may write

$$(a_1, a_2, \ldots, a_n) = (a_1, a_n)(a_1, a_{n-1}) \cdots (a_1, a_2). \tag{8.4}$$

Observe the effect of these two permutations on a_i. For $i \neq 1$ and $i \neq n$, reading from right to left (as the permutation is applied), we see that a_i maps to a_1, which, reading further from right to left, is mapped to a_{i+1}, as we would expect. If $i = 1$, then a_1 maps to a_2, and there is no further mapping. Finally, if $i = n$, then we read left to right to the only transposition containing a_n and see that a_n maps to a_1. Thus, Eq. (8.4) holds. This completes the proof. □

Remark 8.36. The following theorem is useful for our work on matrices in the second part of this chapter, but its proof is outside the scope of these notes. The interested reader can see Chapter 2.2 of Ref. [80].

Theorem 8.37. *No permutation can be expressed as both a product of an even and an odd number of transpositions.* □

Definition 8.38 (Even/Odd permutation). Let $\sigma \in S_n$ be a permutation. If σ can be expressed as an *even* number of transpositions, then it is *even*, otherwise σ is *odd*. The *signature* of the permutation is

$$\operatorname{sgn}(\sigma) = \begin{cases} -1 & \sigma \text{ is odd,} \\ 1 & \sigma \text{ is even.} \end{cases} \tag{8.5}$$

Remark 8.39. Let $G = (V, E)$ be a graph. If $f \in \operatorname{Aut}(G)$, then f is a permutation on the vertices of G. Thus, the graph automorphism group is just a permutation group that respects vertex adjacency.

Example 8.40. Consider the graph K_3, the complete graph on three vertices (see Fig. 8.3(a)). The graph K_3 has six automorphisms, one for each element in S_3, the set of all permutations on three objects. These automorphisms are: (i) the identity automorphism that maps all vertices to themselves, which is the identity permutation e; (ii) the

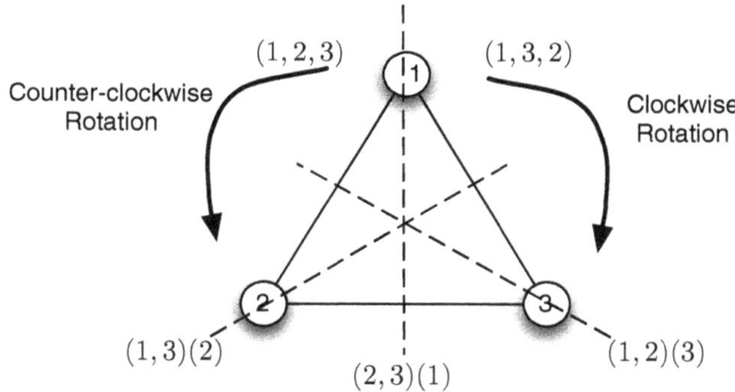

Fig. 8.3 The graph K_3 has six automorphisms, one for each element in S_3, the set of all permutations on three objects. These automorphisms are: (i) the identity automorphism that maps all vertices to themselves; (ii) the automorphism that exchanges vertices 1 and 2; (iii) the automorphism that exchanges vertices 1 and 3; (iv) the automorphism that exchanges vertices 2 and 3; (v) the automorphism that sends 1 to 2, 2 to 3, and 3 to 1; and (vi) the automorphism that sends 1 to 3, 3 to 2, and 2 to 1.

automorphism that exchanges vertices 1 and 2, which is the permutation $(1, 2)$; (iii) the automorphism that exchanges vertices 1 and 3, which is the permutation $(1, 3)$; (iv) the automorphism that exchanges vertices 2 and 3, which is the permutation $(2, 3)$; (v) the automorphism that sends 1 to 2, 2 to 3, and 3 to 1, which is the permutation $(1, 2, 3)$; and (vi) the automorphism that sends 1 to 3, 3 to 2, and 2 to 1, which is the permutation $(1, 3, 2)$.

Note that each of these automorphisms is illustrated by a symmetry in the graphical representation of K_3. The permutations $(1, 2)$, $(1, 3)$, and $(2, 3)$ are flips about an axis of symmetry, while the permutations $(1, 2, 3)$ and $(1, 3, 2)$ are rotations. This is illustrated in Fig. 8.3.

It should be noted that this method of drawing a graph to find its automorphism group does not work in general, but for some graphs (such as complete graphs or cycle graphs), this can be useful.

Lemma 8.41. *The automorphism group of K_n is S_n, thus $|\mathrm{Aut}(K_n)| = n!$* □

Definition 8.42 (Star graph). A *star graph* on $n + 1$ vertices (unfortunately denoted S_n) is a graph with vertex set $V =$

$\{v_0, \ldots, v_n\}$ and edge set E so that

$$e \in E \iff e = \{v_0, v_i\} \quad i \in \{1, \ldots, n\}.$$

Thus, the graph S_n has $n + 1$ vertices and n edges.

Remark 8.43. It is unfortunate that the symmetric group on n items and star graph with $n+1$ vertices have the same representation. We differentiate between the two explicitly to prevent conclusion. It is also worth noting that some references define the star graph S_n to have n vertices and $n - 1$ edges.

Example 8.44. The star graph S_3 with four vertices and three edges is shown in Fig. 8.4, as is the graph S_9.

Remark 8.45. We end this chapter with a simple proposition showing that it is not only complete graphs that can have large automorphism groups.

Proposition 8.46. *The automorphism group of the star graph S_n has $n!$ elements.* □

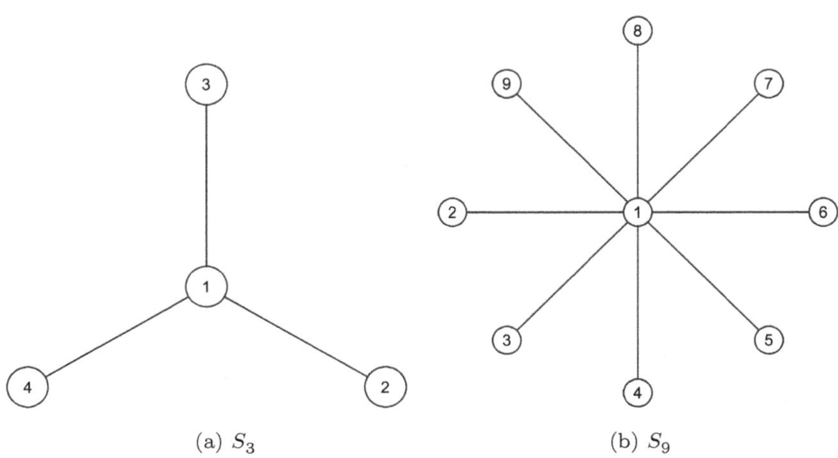

(a) S_3 (b) S_9

Fig. 8.4 The star graphs S_3 and S_9.

8.5 Chapter Notes

In this chapter, we have shown that any graph automorphism is just a permutation on the vertices. Consequently, the identity permutation (mapping vertex v to itself for all v) is always an automorphism of a graph. This is called the trivial automorphism. Determining whether a graph has more automorphisms is called the *graph automorphism problem*. Just as the graph *isomorphism* problem is of unknown complexity, it is also unknown whether the graph automorphism problem is NP-complete [81]. An automorphism $f : V \to V$ has a fixed point if there is a $v \in V$ so that $f(v) = v$. The question of determining whether a graph has an automorphism with no fixed points is known to be NP-complete [81]. The idea of using visual symmetries to find graph automorphisms (as we did for C_3) can be extended [82, 83] to larger graphs, and there are computer programs that try to do this.

It turns out that there is a deeper relationship between graphs and groups. In 1936, König conjectured [84] that every finite group is the automorphism group of some graph. This was proved by Frucht [85] in 1939. It is now known as Frucht's theorem. The proof approach uses a Cayley graph analysis. Cayley graphs (named after Arthur Cayley) encode information about a group's multiplication table. To fully understand them, we need the idea of a generating set of a group. Suppose $(S, \circ)$ is a group. A generating set $H \subset S$ is a subset of S whose elements and their inverses can be used to reconstruct the entire group. For example, $\{1\}$ generates the group of integers under addition because $n = 1+1+\cdots+1$, while $-n = (-1)+(-1)+\cdots+(-1)$, where each sum has n terms. Similarly, the rotations we discussed when building the automorphisms of C_3 are generated by a single $2\pi/3$ radian rotation (which is then applied to itself over and over). The automorphism group of every cycle is generated by a two-element set, and this group is isomorphic to the symmetry group of the regular polygons (called the dihedral group). To construct a Cayley graph:

(1) assign each element in S to a vertex;
(2) assign each element of $h \in H$ a color c_h;
(3) for each $g \in G$ and $h \in H$, add the directed edge (g, gh) with color c_h.

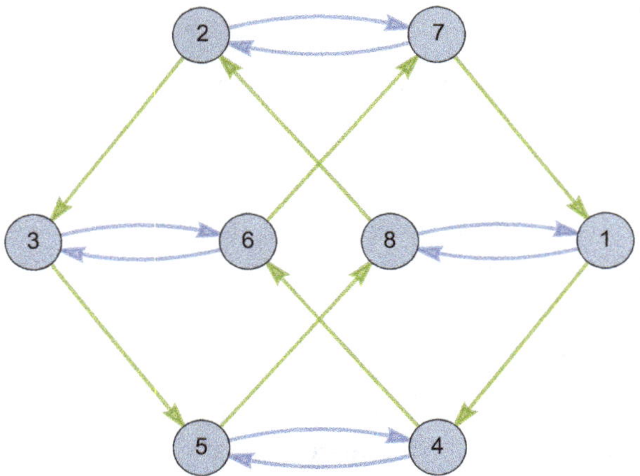

Fig. 8.5 The Cayley graph of $\text{Aut}(C_4)$, which has a generating set of size 2 and eight elements.

The Cayley graph for the group of automorphisms of C_4 is shown in Fig. 8.5. Algebraic concepts also enter into algebraic graph theory through the various graph polynomials that we discussed at the end of Chapter 7. Polynomials (and their roots) have a long and deep connection with group theory. Readers can consult Ref. [80] for details.

8.6 Exercises

Exercise 8.1

Prove that graph isomorphism is an equivalence relation. [Hint: Recall that an equivalence relation is a binary relation $\sim$ defined on a set S so that (i) for all $s \in S$, $s \sim s$ (reflexiveness); (ii) for all $s, t \in S$, $s \sim t \iff t \sim s$ (symmetry); and (iii) for all $r, s, t \in T$, $r \sim s$ and $s \sim t$ implies $r \sim t$ (transitivity). Here, the set is the set of all graphs.]

Exercise 8.2

Prove Theorem 8.12. [Hint: The proof does not have to be extensive in detail. Simply write enough to convince yourself that the isomorphisms preserve the subgraph property.]

Exercise 8.3
List some ways to determine that two graphs are not isomorphic. That is, what are some tests one might do to see whether two graphs are *not* isomorphic?

Exercise 8.4
Prove carefully that if f is a bijection, then so is f^{-1}. [Hint: Most of the proof is in Remark 8.17.]

Exercise 8.5
Prove Lemma 8.20.

Exercise 8.6
Let $(S, \circ)$ be a group with identity e. Prove that the set $\{e\}$ with $\circ$ is also a group called the trivial subgroup of $(S, \circ)$. Conclude that for graphs with only the identity automorphism, their automorphism group is trivial.

Exercise 8.7
Prove Proposition 8.34.

Exercise 8.8
Find the automorphism group of the cycle graph C_4. Can you find the automorphism group for C_k $k \geq 3$? How many elements does it have?

Exercise 8.9
Prove Lemma 8.41.

Exercise 8.10
Show that the automorphism group of the star graph S_3 is also identical to the symmetric permutation group S_3. As a result, show that two non-isomorphic graphs can share an automorphism group. (Remember that $\text{Aut}(K_3)$ is also the symmetric permutation group on three elements.)

Exercise 8.11
Prove Proposition 8.46.

Exercise 8.12

(Project) Study the problem of graph automorphism in detail. Explore the computational complexity of determining the automorphism group of a graph or a family of graphs. Explore any automorphism groups for specific types of graphs, such as cycle graphs, star graphs, and hypercubes.

Chapter 9

Algebraic Graph Theory with Linear Algebra

Remark 9.1 (Chapter goals). Our goal in this chapter is to discuss algebraic graph theory from the perspective of linear algebra. We assume that the reader is familiar with matrix operations, eigenvalues, and eigenvectors. A review of these topics is contained in Appendix A. In this chapter, we discuss the various matrices one can construct from a graph. We study some of the properties of the spectra (eigenvalues) of these matrices. We conclude by introducing the Perron–Frobenius theorem, which we use in the following chapter to discuss applications of algebraic graph theory.

9.1 Matrix Representations of Graphs

Definition 9.2 (Adjacency matrix). Let $G = (V, E)$ be a graph and assume that $V = \{v_1, \ldots, v_n\}$. The *adjacency matrix* of G is an $n \times n$ matrix $\mathbf{M}$ defined as

$$\mathbf{M}_{ij} = \begin{cases} 1 & \{v_i, v_j\} \in E, \\ 0 & \text{otherwise.} \end{cases}$$

Proposition 9.3. *The adjacency matrix of a (simple) graph is symmetric.* □

Theorem 9.4. *Let $G = (V, E)$ be a graph with $V = \{v_1, \ldots, v_n\}$, and let $\mathbf{M}$ be its adjacency matrix. For $k \geq 0$, the (i, j) entry of $\mathbf{M}^k$ is the number of walks of length k from v_i to v_j.*

Proof. We proceed by induction. By definition, $\mathbf{M}^0$ is the $n \times n$ identity matrix, and the number of walks of length 0 between v_i and v_j is 0 if $i \neq j$ and 1 otherwise; thus, the base case is established.

Now, suppose that the (i, j)th entry of $\mathbf{M}^k$ is the number of walks of length k from v_i to v_j. We show that this is true for $k + 1$. We know that

$$\mathbf{M}^{k+1} = \mathbf{M}^k\mathbf{M}. \tag{9.1}$$

Consider vertices v_i and v_j. The (i, j)th element of $\mathbf{M}^{k+1}$ is

$$\mathbf{M}_{ij}^{k+1} = \left(\mathbf{M}_{i\cdot}^{k}\right)\mathbf{M}_{\cdot j}. \tag{9.2}$$

Let

$$\mathbf{M}_{i\cdot}^{k} = \left[r_1 \ldots r_n\right], \tag{9.3}$$

where r_l, $l = 1, \ldots, n$, is the number of walks of length k from v_i to v_l by the induction hypothesis. Let

$$\mathbf{M}_{\cdot j} = \begin{bmatrix} b_1 \\ \vdots \\ b_n \end{bmatrix}, \tag{9.4}$$

where b_l, $l = 1, \ldots, n$, is a 1 if and only if there is an edge $\{v_l, v_j\} \in E$ and 0 otherwise. Then, the $(i, j)^{\text{th}}$ term of $\mathbf{M}^{k+1}$ is

$$\mathbf{M}_{ij}^{k+1} = \mathbf{M}_{i\cdot}^{k}\mathbf{M}_{\cdot j} = \sum_{l=1}^{n} r_l b_l. \tag{9.5}$$

This is the total number of walks of length k leading to a vertex v_l, $l = 1, \ldots, n$, from vertex v_i such that there is also an edge connecting v_l to v_j. Thus, $\mathbf{M}_{ij}^{k+1}$ is the number of walks of length $k + 1$ from v_i to v_j. The result follows by induction. □

Example 9.5. Consider the graph in Fig. 9.1. The adjacency matrix for this graph is

$$\mathbf{M} = \begin{bmatrix} 0 & 1 & 1 & 1 \\ 1 & 0 & 0 & 1 \\ 1 & 0 & 0 & 1 \\ 1 & 1 & 1 & 0 \end{bmatrix}. \tag{9.6}$$

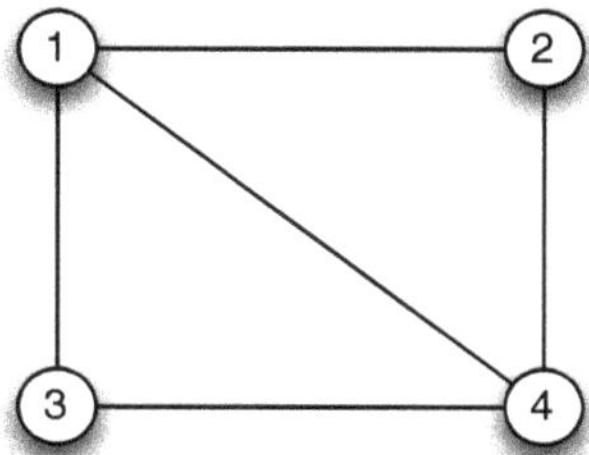

Fig. 9.1 The adjacency matrix of a graph with n vertices is an $n \times n$ matrix with a 1 at element (i, j) if and only if there is an edge connecting vertex i to vertex j; otherwise, element (i, j) is a zero.

Consider $\mathbf{M}^2$:

$$\mathbf{M}^2 = \begin{bmatrix} 3 & 1 & 1 & 2 \\ 1 & 2 & 2 & 1 \\ 1 & 2 & 2 & 1 \\ 2 & 1 & 1 & 3 \end{bmatrix}. \tag{9.7}$$

This tells us that there are three distinct walks of length 2 from vertex v_1 to itself. These walks are

(1) $(v_1, \{v_1, v_2\}, v_2, \{v_1, v_2\}, v_1)$,
(2) $(v_1, \{v_1, v_2\}, v_3, \{v_1, v_3\}, v_1)$, and
(3) $(v_1, \{v_1, v_4\}, v_4, \{v_1, v_4\}, v_1)$.

We also see that there is one path of length 2 from v_1 to v_2: $(v_1, \{v_1, v_4\}, v_4, \{v_2, v_4\}, v_2)$. We can verify each of the other numbers of paths in $\mathbf{M}^2$.

Definition 9.6 (Directed adjacency matrix). Let $G = (V, E)$ be a directed graph, and assume that $V = \{v_1, \dots, v_n\}$. The *adjacency matrix* of G is an $n \times n$ matrix $\mathbf{M}$ defined as

$$\mathbf{M}_{ij} = \begin{cases} 1 & (v_i, v_j) \in E, \\ 0 & \text{otherwise.} \end{cases}$$

Theorem 9.7. *Let $G = (V, E)$ be a digraph with $V = \{v_1, \dots, v_n\}$, and let $\mathbf{M}$ be its adjacency matrix. For $k \geq 0$, the (i, j) entry of $\mathbf{M}^k$ is the number of directed walks of length k from v_i to v_j.* □

Definition 9.8 (Incidence matrix). Let $G = (V, E)$ be a graph with $V = \{v_1, \ldots, v_m\}$ and $E = \{e_1, \ldots, e_n\}$. Then, the *incidence matrix* of G is an $m \times n$ matrix $\mathbf{A}$ with

$$\mathbf{A}_{ij} = \begin{cases} 0 & \text{if } v_i \text{ is not in } e_j, \\ 1 & \text{if } v_i \text{ is in } e_j \text{ and } e_j \text{ is not a self-loop,} \\ 2 & \text{if } v_i \text{ is in } e_j \text{ and } e_j \text{ is a self-loop.} \end{cases} \tag{9.8}$$

Theorem 9.9. *Let $G = (V, E)$ be a graph with $V = \{v_1, \ldots, v_m\}$ and $E = \{e_1, \ldots, e_n\}$ with incidence matrix $\mathbf{A}$. The sum of every column in $\mathbf{A}$ is 2, and the sum of each row in $\mathbf{A}$ is the degree of the vertex corresponding to that row.*

Proof. Consider any column in $\mathbf{A}$; it corresponds to an edge e of G. If the edge is a self-loop, there is only one vertex adjacent to e and thus only one nonzero entry in this column. Therefore, its sum is 2. Conversely, if e connects two vertices, then there are precisely two vertices adjacent to e and thus two entries in this column that are nonzero both with value 1; thus, again, the sum of the column is 2.

Now, consider any row in $\mathbf{A}$; it corresponds to a vertex v of G. The entries in this row are 1 if there is some edge that is adjacent to v and 2 if there is a self-loop at v. From Definition 1.13, we see that adding these values up yields the degree of the vertex v. This completes the proof. □

Definition 9.10 (Directed incidence matrix). Let $G = (V, E)$ be a digraph with $V = \{v_1, \ldots, v_m\}$ and $E = \{e_1, \ldots, e_n\}$. Then, the *incidence matrix* of G is an $m \times n$ matrix $\mathbf{A}$ with

$$\mathbf{A}_{ij} = \begin{cases} 0 & \text{if } v_i \text{ is not in } e_j, \\ 1 & \text{if } v_i \text{ is the source of } e_j \text{ and } e_j \text{ is not a self-loop,} \\ -1 & \text{if } v_i \text{ is the destination of } e_j \text{ and } e_j \text{ is not a self-loop,} \\ 2 & \text{if } v_i \text{ is in } e_j \text{ and } e_j \text{ is a self-loop.} \end{cases} \tag{9.9}$$

Remark 9.11. The adjacency matrices of simple directed graphs (those with no self-loops) have very useful properties, which are discussed in Chapter 12. In particular, these matrices have the property that every square sub-matrix has a determinant that is either 1, −1, or 0. This property is called total unimodularity, and it is particularly important in the analysis of network flows.

9.2 Properties of the Eigenvalues of the Adjacency Matrix

Remark 9.12. Going forward, we assume that the reader is familiar with basic facts about matrix eigenvalues. This information can be found in Appendix A.

Lemma 9.13 (Rational root theorem). Let $a_n x^n + \cdots + a_1 x + a_0 = 0$ for $x = p/q$ with $\gcd(p, q) = 1$ and $a_n, \ldots, a_0 \in \mathbb{Z}$. This is p/q is a rational root of the equation. Then, p is an integer factor of a_0 and q is an integer factor of a_n. □

Remark 9.14. The following theorem follows from the spectral theorem for real symmetric matrices (see Theorem A.97) and the rational root theorem.

Theorem 9.15 (Graph spectrum). *Let $G = (V, E)$ be a graph with adjacency matrix $\mathbf{M}$. The following hold:*

(1) *Every eigenvalue of $\mathbf{M}$ is real.*
(2) *If λ is a rational eigenvalue of $\mathbf{M}$, then it is an integer.* □

Remark 9.16. Two graphs that are not isomorphic can have the same set of eigenvalues. This can be illustrated through an example that can be found in Chapter 8 of Ref. [39]. The graphs are shown in Fig. 9.2. We can see that the two graphs are not isomorphic since there is no vertex in Graph G_1 that has a degree of 6 unlike Vertex 7 of graph G_2. However, one can determine (using a computer) that their adjacency matrices share the same set of eigenvalues.

Definition 9.17 (Irreducible matrix). A matrix $\mathbf{M} \in \mathbb{R}^{n\times n}$ is *irreducible* if for each pair (i, j), there is some $k \in \mathbb{Z}$ with $k > 0$ so that $\mathbf{M}_{ij}^k > 0$.

Lemma 9.18. *If $G = (V, E)$ is a connected graph with adjacency matrix $\mathbf{M}$, then $\mathbf{M}$ is irreducible.* □

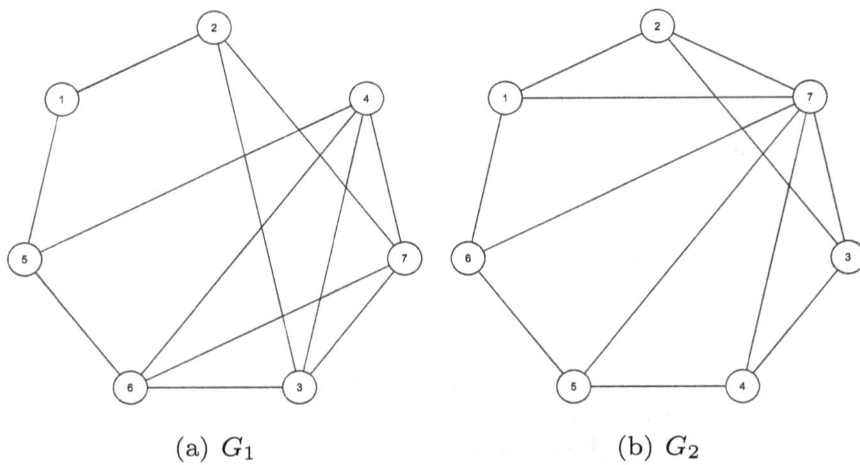

Fig. 9.2 Two graphs with the same eigenvalues that are not isomorphic are illustrated.

Theorem 9.19 (Perron–Frobenius theorem). *If* $\mathbf{M}$ *is an irreducible matrix, then* $\mathbf{M}$ *has an eigenvalue of* λ_0 *with the following properties*:

(1) *The eigenvalue* λ_0 *is positive, and if* λ *is an alternative eigenvalue of* $\mathbf{M}$, *then* $\lambda_0 \geq |\lambda|$.
(2) *The matrix* $\mathbf{M}$ *has an eigenvector* $\mathbf{v}_0$ *corresponding to* λ_0 *with only positive entries when properly scaled.*
(3) *The eigenvalue* λ_0 *is a simple root of the characteristic equation for* $\mathbf{M}$ *and, therefore, has a unique (up to scale) eigenvector* $\mathbf{v}_0$.
(4) *The eigenvector* $\mathbf{v}_0$ *is the only eigenvector of* $\mathbf{M}$ *that can have all positive entries when properly scaled.* □

Corollary 9.20. *If* $G = (V, E)$ *is a connected graph with the adjacency matrix* $\mathbf{M}$, *then it has a unique largest eigenvalue that corresponds to an eigenvector that is positive when properly scaled.*

Proof. Applying Lemma 9.18, we see that $\mathbf{M}$ is irreducible. Furthermore, we know that there is an eigenvalue λ_0 of $\mathbf{M}$ that is (i) greater than or equal to, in absolute value, all other eigenvalues of $\mathbf{M}$ and (ii) a simple root. From Theorem 9.15, we know that all eigenvalues of $\mathbf{M}$ are real. But for (i) and (ii) to hold, no other (real) eigenvalue can have value equal to λ_0 (otherwise, it would not be a simple root). Thus, λ_0 is the unique largest eigenvalue of $\mathbf{M}$. This completes the proof. □

9.3 Chapter Notes

Matrices (in the form used in this chapter) are a relatively recent invention, though arrays of numbers have been used for centuries. The term "matrix" was first used by J. J. Sylvester in Ref. [86]. At this time, matrices were mainly used as generators of determinants. Cayley (of Cayley graphs from the notes in Chapter 8) used matrices in his work on geometric transformations [87] and proved the basic algebraic properties of matrix arithmetic. He later wrote a treatise on matrices [88]. According to Ref. [87], Cullis was the first to use the square-bracket notation in 1913 [89].

The adjacency matrix of a graph was known and used at least by Pólya and Szegő [90] in 1925. This fact is mentioned by Harary [91] in his 1962 paper on determinants of adjacency matrices. In particular, Harary states that Pólya suggested the paper to him. The question of non-isomorphic graphs that share spectra (adjacency matrix eigenvalues) is one considered in Ref. [91]. Questions on the properties of the spectra of graphs are also studied in Refs. [92, 93]. The incidence matrix, which we did not discuss in detail, returns in Chapter 12, where it appears as the constraint matrix of the linear programming problem that arises from network flow problems.

Oskar Perron was not a graph theorist. He was a professor of mathematics at the University of Heidelberg, where he studied ordinary and partial differential equations. Likewise, Frobenius was a mathematician who studied differential equations (as well as number and group theory). The Perron–Frobenius theorem was proved independently by the two mathematicians. This result has found applications ranging from graph theory (see the following chapter) to economics [94] and even ranking football (soccer) teams [95, 96].

The Perron–Frobenius theorem is a classical result in linear algebra with several proofs (see Ref. [97]). Meyer says the following about the theorem:

> *In addition to saying something useful, the Perron–Frobenius theory is elegant. It is a testament to the fact that beautiful mathematics eventually tends to be useful, and useful mathematics eventually tends to be beautiful.*

One should note that you can say more than we state in our presentation of the theorem. See Chapter 8 of Ref. [97] for details and a proof. In the next chapter, we discuss several applications of the Perron–Frobenius theorem arising from graphs.

9.4 Exercises

Exercise 9.1
Find the adjacency matrix for the graph C_4. Can you describe the adjacency matrix for an arbitrary cycle C_k?

Exercise 9.2
Prove Proposition 9.3.

Exercise 9.3
Devise an inefficient test for isomorphism between two graphs G and G' using their adjacency matrix representations. Assume that it takes 1 time unit to test whether two $n \times n$ matrices are equal. What is the maximum amount of time your algorithm takes to determine that $G \ncong G'$? [Hint: Continue to reorder the vertices of G' and test the adjacency matrices for equality.]

Exercise 9.4
Prove Theorem 9.7. [Hint: Use the approach in the proof of Theorem 9.4.]

Exercise 9.5
Use Theorem 9.9 to prove Theorem 2.10 a new way.

Exercise 9.6
(**Project**) Prove the spectral theorem for real symmetric matrices and then use it to obtain Part 1 of Theorem 9.15. Then, prove and apply Lemma 9.13 to prove Part 2 of Theorem 9.15. You should discuss the proof of the spectral theorem for real symmetric matrices. [Hint: All these proofs are available in references or online; expand on these sources in your own words.]

Exercise 9.7
Use a computer to show that the two graphs in Remark 9.16 share the same set of eigenvalues.

Exercise 9.8
Prove Lemma 9.18.

Exercise 9.9
Find the principal eigenvector/eigenvalue pair of the adjacency matrix for S_4, the star graph with five vertices. By principal eigenvector/eigenvalue pair, we mean the eigenvalue with the largest positive value and its corresponding eigenvector.

Chapter 10

Applications of Algebraic Graph Theory

Remark 10.1 (Chapter goals). In this chapter, we explore applications of algebraic graph theory. We study additional network centrality measures, explore Markov chains (a practical application of directed graphs), and study spectral clustering of graphs. The first two topics are motivated by the Perron–Frobenius theorem, while the last topic is a direct application of the graph Laplacian matrix.

10.1 Eigenvector Centrality

Remark 10.2. The following approach to deriving eigenvector centrality comes from Leo Spizzirri [98]. What follows is a derivation, not a proof.

Derivation 10.3 (Eigenvector centrality). We can assign to each vertex of a graph $G = (V, E)$ a score (called its eigenvector centrality) that will determine its *relative importance* in the graph. Here, importance is measured in a self-referential way: Important vertices are important precisely because they are adjacent to other important vertices. (This is the high-school concept of "coolness by association.") This self-referential definition can be resolved in the following way.

Let x_i be the (unknown) score of vertex $v_i \in V$, and let $x_i = \kappa(v_i)$, with κ being the function returning the score of each vertex in V. Define x_i as a pseudo-average of the scores of its neighbors.

That is, write

$$x_i = \frac{1}{\lambda} \sum_{v \in N(v_i)} \kappa(v). \tag{10.1}$$

Here, λ will be chosen endogenously during computation.

Recall that $\mathbf{M}_{i\cdot}$ is the ith row of the adjacency matrix $\mathbf{M}$ and contains a 1 in position j if and only if v_i is adjacent to v_j, i.e., to say $v_j \in N(v_i)$. Thus, we can rewrite Eq. (10.1) as

$$x_i = \frac{1}{\lambda} \sum_{j=1}^{n} \mathbf{M}_{ij} x_j.$$

This leads to n equations, one for each vertex in V (or each row of $\mathbf{M}$). Written as a matrix expression, we have

$$\mathbf{x} = \frac{1}{\lambda}\mathbf{M}\mathbf{x} \implies \lambda\mathbf{x} = \mathbf{M}\mathbf{x}. \tag{10.2}$$

Thus, $\mathbf{x}$ is an eigenvector of $\mathbf{M}$ and λ is its eigenvalue.

Remark 10.4. Clearly, there may be several eigenvectors and eigenvalues for $\mathbf{M}$. The question is, which eigenvalue–eigenvector pair should be chosen? The answer is to choose the eigenvector with all positive entries corresponding to the largest eigenvalue. We know such an eigenvalue–eigenvector pair exists and is unique as a result of Lemma 9.18 and the Perron–Frobenius theorem (Theorem 9.19).

Theorem 10.5. *Let $G = (V, E)$ be a connected graph with adjacency matrix $\mathbf{M} \in \mathbb{R}^{n\times n}$. Suppose that λ_0 is the largest real eigenvalue of $\mathbf{M}$ and has the corresponding eigenvector $\mathbf{v}_0$. Furthermore, assume that $|\lambda_0| > |\lambda|$ for any other eigenvalue λ of M. If $\mathbf{x} \in \mathbb{R}^{n\times 1}$ is a column vector so that $\mathbf{x} \cdot \mathbf{v_0} \neq 0$, then*

$$\lim_{k\to\infty} \frac{\mathbf{M}^k\mathbf{x}}{\lambda_0^k} = \alpha_0\mathbf{v_0}. \tag{10.3}$$

Proof. Applying Theorem A.97, we see that the eigenvectors of $\mathbf{M}$ must form a basis for $\mathbb{R}^n$. Thus, we can express

$$\mathbf{x} = \alpha_0\mathbf{v}_0 + \alpha_1\mathbf{v}_1 + \cdots + \alpha_{n-1}\mathbf{v}_{n-1}. \tag{10.4}$$

Multiplying both sides by $\mathbf{M}^k$ yields

$$\begin{aligned}\mathbf{M}^k\mathbf{x} &= \alpha_0\mathbf{M}^k\mathbf{v}_0 + \alpha_1\mathbf{M}^k\mathbf{v}_1 + \cdots + \alpha_{n-1}\mathbf{M}^k\mathbf{v}_{n-1}\\ &= \alpha_0\lambda_0^k\mathbf{v}_0 + \alpha_1\lambda_1^k\mathbf{v}_1 + \cdots + \alpha_{n-1}\lambda_n^k\mathbf{v}_{n-1} \qquad (10.5)\end{aligned}$$

because $\mathbf{M}^k\mathbf{v}_i = \lambda_i^k\mathbf{v}_i$ for any eigenvalue $\mathbf{v}_i$. Dividing by λ_0^k yields

$$\frac{\mathbf{M}^k\mathbf{x}}{\lambda_0^k} = \alpha_0\mathbf{v}_0 + \alpha_1\frac{\lambda_1^k}{\lambda_0^k}\mathbf{v}_1 + \cdots + \alpha_{n-1}\frac{\lambda_{n-1}^k}{\lambda_0^k}\mathbf{v}_{n-1}. \qquad (10.6)$$

Applying our assumption that $\lambda_0 > |\lambda|$ for all other eigenvalues λ, we have

$$\lim_{k\to\infty}\frac{\lambda_i^k}{\lambda_0^k} = 0 \qquad (10.7)$$

for $i \neq 0$. Thus,

$$\lim_{k\to\infty}\frac{\mathbf{M}^k\mathbf{x}}{\lambda_0^k} = \alpha_0\mathbf{v_0}. \qquad (10.8)$$

□

Remark 10.6. We can use Theorem 10.5 to justify our definition of eigenvector centrality as the Perron–Frobenius eigenvector of the adjacency matrix. Let $\mathbf{x}$ be a vector with a 1 at index i and 0 everywhere else. We imagine this vector corresponds to a walker who starts at vertex v_i in the graph G. If $\mathbf{M}$ is the adjacency matrix, then $\mathbf{Mx}$ is the ith column of $\mathbf{M}$, whose jth index tells us the number of walks of length 1 leading from vertex v_j to vertex v_i. We can repeat this logic to see that $\mathbf{M}^k\mathbf{x}$ gives us a vector whose jth element is the number of walks of length k from v_j to v_i.

From Theorem 10.5, we know that (under some suitable conditions), no matter which vertex we choose in creating $\mathbf{x}$,

$$\lim_{k\to\infty}\frac{\mathbf{M}^k\mathbf{x}}{\lambda_0^k} = \alpha_0\mathbf{v_0}. \qquad (10.9)$$

Reinterpreting Eq. (10.9), we observe that as $k \to \infty$, the eigenvector centrality of vertex v_i is just counting (a scaled version) of the number of long paths leading to that vertex. The more paths leading to a vertex, the more central it is and thus the higher it is ranked by eigenvector centrality.

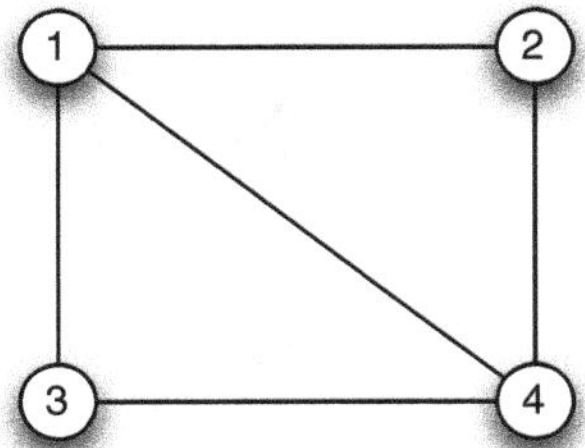

Fig. 10.1 A matrix with four vertices and five edges. Intuitively, vertices 1 and 4 should have the same eigenvector centrality score as vertices 2 and 3.

Example 10.7. Consider the graph shown in Fig. 10.1. Recall from Example 9.5 that this graph has the adjacency matrix

$$\mathbf{M} = \begin{bmatrix} 0 & 1 & 1 & 1 \\ 1 & 0 & 0 & 1 \\ 1 & 0 & 0 & 1 \\ 1 & 1 & 1 & 0 \end{bmatrix}.$$

We can use a computer to determine the eigenvalues and eigenvectors of $\mathbf{M}$. The eigenvalues are

$$\left\{\frac{1}{2}\left(1+\sqrt{17}\right), \frac{1}{2}\left(1-\sqrt{17}\right), -1, 0\right\},$$

while the corresponding eigenvectors are the rows of the matrix

$$\begin{bmatrix} 1 & -\frac{-3-\sqrt{17}}{5+\sqrt{17}} & -\frac{-3-\sqrt{17}}{5+\sqrt{17}} & 1 \\ 1 & -\frac{3-\sqrt{17}}{\sqrt{17}-5} & -\frac{3-\sqrt{17}}{\sqrt{17}-5} & 1 \\ -1 & 0 & 0 & 1 \\ 0 & -1 & 1 & 0 \end{bmatrix}.$$

The largest eigenvalue is $\lambda_0 = \frac{1}{2}\left(1+\sqrt{17}\right)$, which has the corresponding eigenvector

$$\mathbf{v}_0 = \begin{bmatrix} 1 & -\dfrac{-3-\sqrt{17}}{5+\sqrt{17}} & -\dfrac{-3-\sqrt{17}}{5+\sqrt{17}} & 1 \end{bmatrix}.$$

Eigenvector centrality is usually normalized, so the entries of the vector sum to one. This can be accomplished as

$$\hat{\mathbf{v}}_0 = \frac{\mathbf{v}_0}{\sum_j \mathbf{v}_{0_j}}.$$

In this example, our normalized eigenvector centrality is given by the approximation

$$\mathbf{v}_0 \approx \begin{bmatrix} 0.28 & 0.22 & 0.22 & 0.28 \end{bmatrix}.$$

This illustrates more clearly that vertices 1 and 4 have identical (larger) eigenvector centrality scores and vertices 2 and 3 have identical (smaller) eigenvector centrality scores. Let $\mathbf{x} = \langle 1, 0, 0, 0 \rangle$ and $\mathbf{y}_k = \mathbf{M}^k\mathbf{x}$, and suppose that $\hat{\mathbf{y}}_k$ is normalized, so its components sum to one. Then,

$$\hat{\mathbf{y}}_1 \approx \begin{bmatrix} 0 \\ 0.33 \\ 0.33 \\ 0.33 \end{bmatrix} \quad \hat{\mathbf{y}}_5 \approx \begin{bmatrix} 0.26 \\ 0.24 \\ 0.24 \\ 0.27 \end{bmatrix} \quad \hat{\mathbf{y}}_{10} \approx \begin{bmatrix} 0.28 \\ 0.22 \\ 0.22 \\ 0.28 \end{bmatrix}.$$

It's easy to see that as $k \to \infty$, $\hat{\mathbf{y}}_k$ approaches $\hat{\mathbf{v}}_0$ as expected.

10.2 Markov Chains and Random Walks

Remark 10.8. Appendix B provides an introduction to probability. While it is not necessary, it may make some of the following observations on Markov chains easier to understand for those with no background in probability theory.

Definition 10.9 (Markov chain). A *discrete-time Markov chain* is a pair $\mathcal{M} = (G, p)$ where $G = (V, E)$ is a *directed* graph and the set of vertices is usually called the set of *states*, the set of edges are called the *transitions*, and $p : E \to [0, 1]$ is a probability assignment function satisfying

$$\sum_{v' \in N_o(v)} p(v, v') = 1, \tag{10.10}$$

for all $v \in V$. Here, $N_o(v)$ is the neighborhood reachable by an out-edge from v. If there is no edge $(v, v') \in E$, then $p(v, v') = 0$.

Remark 10.10. There are also continuous-time Markov chains, but we will not discuss those here. See Ref. [99] for information on those models. For the remainder of this chapter, when we say Markov chain, we mean discrete-time Markov chain.

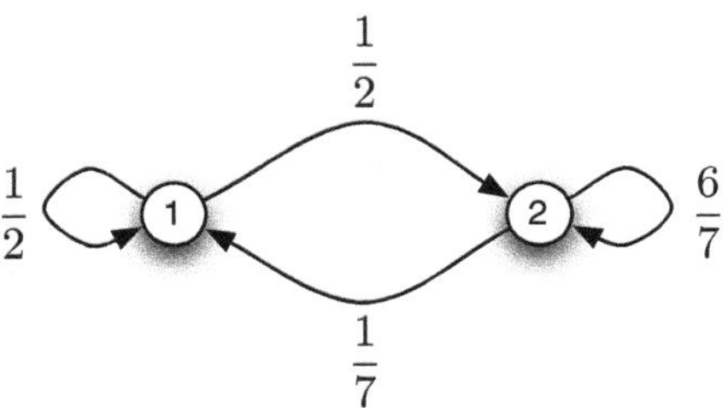

Fig. 10.2 A Markov chain is a directed graph to which we assign edge probabilities so that the sum of the probabilities of the out-edges at any vertex is always 1.

Example 10.11. A simple Markov chain is shown in Fig. 10.2. There are two states (vertices) denoted by 1 and 2. The fractions next to the directed edges are their assigned probabilities. We can think of a Markov chain as governing the evolution of a state as follows. Think of the states as cities with airports. If there is an out-edge connecting the current city to another city, then we can fly from our current city to this next city, and we do so with some probability. When we do fly (or perhaps don't fly and remain at the current location), our state updates to the next city. In this case, time is treated *discretely.*

A walk along the vertices of a Markov chain governed by the probability function is called a *random walk.* Computing the probability of taking a specific random walk is an exercise in conditional probability (see Appendix B).

Definition 10.12 (Stochastic matrix). Let $\mathcal{M} = (G, p)$ be a Markov chain. Then, the *stochastic matrix* (or probability transition matrix) of $\mathcal{M}$ is

$$\mathbf{M}_{ij} = p(v_i, v_j). \tag{10.11}$$

Example 10.13. The stochastic matrix for the Markov chain in Fig. 10.2 is

$$\mathbf{M} = \begin{bmatrix} \frac{1}{2} & \frac{1}{2} \\ \frac{1}{7} & \frac{6}{7} \end{bmatrix}.$$

Thus, a stochastic matrix is very much like an adjacency matrix where the 0's and 1's, indicating the presence and absence of an edge, respectively, are replaced by the probabilities associated with the edges in the Markov chain.

Definition 10.14 (State probability vector). If $\mathcal{M} = (G, p)$ is a Markov chain with n states (vertices), then a *state probability vector* is a vector $\mathbf{x} \in \mathbb{R}^{n\times 1}$ such that $x_1 + x_2 + \cdots + x_n = 1$ and $x_i \geq 0$ for $i = 1, \ldots, n$ and x_i represents the probability that we are in state i (at vertex i).

Remark 10.15. The following theorem can be proved in exactly the same way that Theorem 9.4 is proved.

Theorem 10.16. *Let $\mathcal{M} = (G, p)$ be a Markov chain with n states (vertices). Let $\mathbf{x}^{(0)} \in \mathbb{R}^{n\times 1}$ be an (initial) state probability vector. Then, assuming we take a random walk of length k in $\mathcal{M}$ using the initial state probability vector $\mathbf{x}^{(0)}$, the final state probability vector is*

$$\mathbf{x}^{(k)} = \left(\mathbf{M}^T\right)^k \mathbf{x}^{(0)}. \tag{10.12}$$

□

Remark 10.17. This can be written without the transpose. Let $\mathbf{x}^{(0)} \in \mathbb{R}^{1\times n}$; that is, $\mathbf{x}^{(0)}$ is a row vector. Then,

$$\mathbf{x}^{(k)} = \mathbf{x}^{(0)}\mathbf{M}^k, \tag{10.13}$$

with $\mathbf{x}^{(k)} \in \mathbb{R}^{1\times n}$.

Example 10.18. Consider the Markov chain in Fig. 10.2. The state vector

$$\mathbf{x}^{(0)} = \begin{bmatrix} 1 \\ 0 \end{bmatrix}$$

states that we start in State 1 with probability 1. From Example 10.13, we know what $\mathbf{M}$ is. Then, it is easy to see that

$$\mathbf{x}^{(1)} = \left(\mathbf{M}^T\right)^k \mathbf{x}^{(0)} = \begin{bmatrix} \frac{1}{2} \\ \frac{1}{2} \end{bmatrix}.$$

This is precisely the state probability vector we would expect after a random walk of length 1 in $\mathcal{M}$.

Definition 10.19 (Stationary probability vector). Let $\mathcal{M} = (G, p)$ be a Markov chain. Then, a vector $\mathbf{x}^*$ is stationary for $\mathcal{M}$ if

$$\mathbf{x}^* = \mathbf{M}^T\mathbf{x}^*. \tag{10.14}$$

Remark 10.20. Equation (10.14) says that a probability distribution on the states of a Markov chain is stationary if it doesn't change as a result of taking a random step in a Markov chain.

Remark 10.21. Equation (10.14) should look familiar. It says that $\mathbf{M}^T$ has an eigenvalue of 1 and a corresponding eigenvector whose entries are all non-negative (so that the vector can be scaled so its components sum to 1). Furthermore, this looks very similar to the equation we used for eigenvector centrality.

Lemma 10.22. *Let $\mathcal{M} = (G, p)$ be a Markov chain with n states and with stochastic matrix $\mathbf{M}$. Then,*

$$\sum_j \mathbf{M}_{ij} = 1 \tag{10.15}$$

for all $i = 1, \ldots, n$. □

Lemma 10.23. *$\mathcal{M} = (G, p)$ be a Markov chain with n states and with the stochastic matrix $\mathbf{M}$. If G is strongly connected, then $\mathbf{M}$ and $\mathbf{M}^T$ are irreducible.*

Proof. If G is strongly connected, then there is a directed walk from any vertex v_i to any other vertex v_j in V, the vertex set of G. Consider any length-k walk connecting v_i to v_j (such a walk exists for some k). Let $\mathbf{e}_i$ be a vector with 1 in its ith component and 0 everywhere else. Then, $(\mathbf{M}^T)^k\mathbf{e}_i$ is the final state probability vector associated with a walk of length k starting at vertex v_i. Since there is a walk of length k from v_i to v_j, we know that the jth element of this vector must be nonzero. That is,

$$\mathbf{e}_j^T(\mathbf{M}^T)^k\mathbf{e}_i > 0,$$

where $\mathbf{e}_j$ is defined just as $\mathbf{e}_i$ is but with 1 at the jth position. Thus, $(\mathbf{M}^T)^k_{ij} > 0$ for some k for every (i, j) pair, and thus, $\mathbf{M}^T$ is irreducible. The fact that $\mathbf{M}$ is irreducible follows immediately from the fact that $(\mathbf{M}^T)^k = (\mathbf{M}^k)^T$. This completes the proof. □

Theorem 10.24 (Perron–Frobenius theorem redux). *If $\mathbf{M}$ is an irreducible matrix, then $\mathbf{M}$ has an eigenvalue $\lambda_0 > 0$ that satisfies all the properties in Theorem 9.19 and*

$$\min_i \sum_j \mathbf{M}_{ij} \leq \lambda_0 \leq \max_i \sum_j \mathbf{M}_{ij}.$$

□

Theorem 10.25. *Let $\mathcal{M} = (G, p)$ be a Markov chain with the stochastic matrix* $\mathbf{M}$. *If* $\mathbf{M}^T$ *is irreducible, then $\mathcal{M}$ has a unique stationary probability distribution.*

Proof. From Theorem A.80, we know that $\mathbf{M}$ and $\mathbf{M}^T$ have identical eigenvalues. By the Perron–Frobenius theorem, $\mathbf{M}$ has the largest positive eigenvalue λ_0 that satisfies

$$\min_i \sum_j \mathbf{M}_{ij} \leq \lambda_0 \leq \max_i \sum_j \mathbf{M}_{ij}.$$

By Lemma 10.22, we know that

$$\min_i \sum_j \mathbf{M}_{ij} = \max_i \sum_j \mathbf{M}_{ij} = 1.$$

Therefore, by the squeezing lemma, $\lambda_0 = 1$. The fact that $\mathbf{M}^T$ has exactly one strictly positive eigenvector $\mathbf{v}_0$ corresponding to $\lambda_0 = 1$ means that

$$\mathbf{M}^T \mathbf{v}_0 = \mathbf{v}_0. \tag{10.16}$$

Thus, $\mathbf{v}_0$ is the unique stationary state probability vector for $\mathcal{M} = (G, p)$. This completes the proof. □

Remark 10.26. Let $\mathcal{M}$ be a Markov chain. If $\lambda_0 = 1$ has a strictly greater absolute value than all other eigenvalues of $\mathbf{M}^T$, then we can strengthen Theorem 10.25 to say that $\lim_{k\to\infty} \mathbf{M}^k$ converges to a rank 1 matrix whose rows are all the stationary distribution vector. This is not true of all Markov chains (because of the eigenvalue requirement). Markov chains for which this property holds may be called *ergodic*; they are *aperiodic*, meaning that a state pattern does not repeat in a cycle and their directed graphs are strongly connected.

10.3 PageRank

Definition 10.27 (Induced Markov chain). Let $G = (V, E)$ be a graph. Then, the *Markov chain induced* from G is the one obtained by defining a new directed graph $G' = (V, E')$, with each edge

$\{v, v'\} \in E$ replaced by two directional edges (v, v') and (v', v) in E and defining the probability function p so that

$$p(v, v') = \frac{1}{\deg_{out_{G'}} v}. \tag{10.17}$$

Example 10.28. An induced Markov chain is shown in Fig. 10.3. The Markov chain in the figure has the stationary state probability vector

$$\mathbf{x}^* = \begin{bmatrix} \frac{3}{8} \\ \frac{2}{8} \\ \frac{2}{8} \\ \frac{1}{8}, \end{bmatrix},$$

which is the eigenvector corresponding to the eigenvalue 1 in the matrix $\mathbf{M}^T$. Arguing as we did in the proof of Theorem 10.5 and Example 10.7, we could expect that for any state vector $\mathbf{x}$, we would have

$$\lim_{k \to \infty} \left(\mathbf{M}^T\right)^k \mathbf{x} = \mathbf{x}^*.$$

We would be correct. When this convergence happens *quickly* (where we leave *quickly* poorly defined), the graph is said to have a *fast mixing* property.

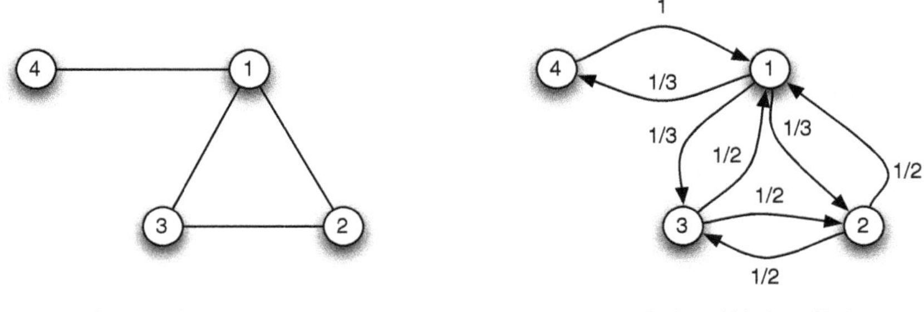

Fig. 10.3 An induced Markov chain is constructed from a graph by replacing every edge with a pair of directed edges (going in opposite directions) and assigning a probability equal to the out-degree of each vertex to every edge leaving that vertex.

If we used the stationary probability of a vertex in the induced Markov chain as a measure of importance, then, clearly, vertex 1 would be the most important, followed by vertices 2 and 3 and, lastly, vertex 4. We can compare this with the eigenvector centrality measure, which assigns a rank vector of

$$\mathbf{x}^+ \approx \begin{bmatrix} 0.315 \\ 0.270 \\ 0.270 \\ 0.145 \end{bmatrix}.$$

The eigenvector centrality gives the same ordinal ranking as using the stationary state probability vector, but there are subtle differences in the values produced by these two ranking schemes. This leads us to PageRank [27].

Derivation 10.29 (PageRank). Consider a collection of web pages each with links. We can construct a directed graph G with the vertex set V consisting of the web pages and edge set E consisting of the directed links among the pages. Imagine a random web surfer who will click among these web pages, following links until a dead end is reached (a page with no outbound links). In this case, the web surfer will type a new URL in (chosen from the set of web pages available) and the process will continue.

From this model, we can induce a Markov chain in which we define a new graph G' with edge set E' so that if $v \in V$ has an out-degree of 0, then we create an edge in E' to every other vertex in V, and we then define

$$p(v, v') = \frac{1}{\deg_{out_{G'}} v} \tag{10.18}$$

exactly as before. In the absence of *any further* insight, the PageRank algorithm simply assigns to each web page a score equal to the stationary probability of the corresponding state in the induced Markov chain. For the remainder of this derivation, let $\mathbf{M}$ be the stochastic matrix of the induced Markov chain.

PageRank assumes that surfers will get bored after some number of clicks (or new URLs) and will stop (and move to a new page) with some probability $d \in [0, 1]$ called the damping factor. This factor is usually estimated. Assuming that there are n web pages, let $\mathbf{r} \in \mathbb{R}^{n\times 1}$ be the PageRank score for each page. Taking boredom into account leads to a new expression for rank (similar to Eq. (10.1) for eigenvector centrality):

$$r_i = \frac{1-d}{n} + d\left(\sum_{j=1}^{n} \mathbf{M}_{ji} r_j\right) \quad \text{for } i = 1, \ldots, n. \tag{10.19}$$

Here, the d term acts like a damping factor on walks through the Markov chain. In essence, it stalls people as they walk, making it less likely that a searcher will keep walking forever. The original system of equations in Eq. (10.19) can be written in matrix form as

$$\mathbf{r} = \left(\frac{1-d}{n}\right)\mathbf{1} + d\mathbf{M}^T\mathbf{r}, \tag{10.20}$$

where $\mathbf{1}$ is a $n \times 1$ vector consisting of all 1's. It is easy to see that when $d = 1$, $\mathbf{r}$ is precisely the stationary state probability vector for the induced Markov chain. When $d \neq 1$, $\mathbf{r}$ is usually computed iteratively by starting with an initial value of $r_i^0 = 1/n$ for all $i = 1, \ldots, n$ and computing

$$\mathbf{r}^{(k)} = \left(\frac{1-d}{n}\right)\mathbf{1} + d\mathbf{M}^T\mathbf{r}^{(k-1)}.$$

The reason is that for large n, the analytic solution

$$\mathbf{r} = \left(\mathbf{I}_n - d\mathbf{M}^T\right)^{-1}\left(\frac{1-d}{n}\right)\mathbf{1} \tag{10.21}$$

is not computationally tractable.[1]

Example 10.30. Consider the induced Markov chain in Fig. 10.3, and suppose we wish to compute PageRank on these vertices with

[1]Note that $\left(\mathbf{I}_n - d\mathbf{M}^T\right)^{-1}$ computes a matrix inverse. We should note that for stochastic matrices, this inverse is guaranteed to exist. For those interested, please consult Refs. [97, 100, 101].

$d = 0.85$ (which is a common assumption). We might begin with

$$\mathbf{r}^{(0)} = \begin{bmatrix} \frac{1}{4} \\ \frac{1}{4} \\ \frac{1}{4} \\ \frac{1}{4} \end{bmatrix}.$$

We would then compute

$$\mathbf{r}^{(1)} = \left(\frac{1-d}{n}\right)\mathbf{1} + d\mathbf{M}^T\mathbf{r}^{(0)} \approx \begin{bmatrix} 0.4625 \\ 0.2146 \\ 0.2146 \\ 0.1083 \end{bmatrix}.$$

We would repeat this again to obtain

$$\mathbf{r}^{(2)} = \left(\frac{1-d}{n}\right)\mathbf{1} + d\mathbf{M}^T\mathbf{r}^{(1)} \approx \begin{bmatrix} 0.312 \\ 0.260 \\ 0.260 \\ 0.160 \end{bmatrix}.$$

This would continue until the difference between the values of $\mathbf{r}^{(k)}$ and $r^{(k-1)}$ was small. The final solution would be close to the exact solution:

$$\mathbf{r}^* \approx \begin{bmatrix} 0.367 \\ 0.246 \\ 0.246 \\ 0.141 \end{bmatrix}.$$

Note this is (again) very close to the stationary probabilities and the eigenvector centralities we observed earlier. This vector is normalized so that all the entries sum to 1.

10.4 The Graph Laplacian

Remark 10.31. In this last section, we return to simple graphs and discuss the *Laplacian matrix*, which can be used to partition the vertices of a graph in a sensible way.

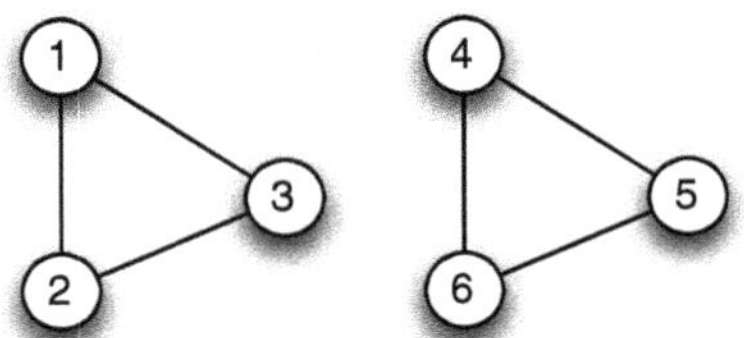

Fig. 10.4 A set of triangle graphs.

Definition 10.32 (Degree matrix). Let $G = (V, E)$ be a simple graph with $V = \{v_1, \ldots, v_n\}$. The *degree matrix* is the diagonal matrix $\mathbf{D}$ with the degree of each vertex on the diagonal. That is, $\mathbf{D}_{ii} = \deg(v_i)$ and $\mathbf{D}_{ij} = 0$ if $i \neq j$.

Example 10.33. Consider the graph in Fig. 10.4. It has the degree matrix

$$\mathbf{D} = \begin{bmatrix} 2 & 0 & 0 & 0 & 0 & 0 \\ 0 & 2 & 0 & 0 & 0 & 0 \\ 0 & 0 & 2 & 0 & 0 & 0 \\ 0 & 0 & 0 & 2 & 0 & 0 \\ 0 & 0 & 0 & 0 & 2 & 0 \\ 0 & 0 & 0 & 0 & 0 & 2 \end{bmatrix},$$

because each of its vertices has a degree of 2.

Definition 10.34 (Laplacian matrix). Let $G = (V, E)$ be a simple graph with $V = \{v_1, \ldots, v_n\}$, adjacency matrix $\mathbf{M}$m and degree matrix $\mathbf{D}$. The *Laplacian matrix* is the matrix $\mathbf{L} = \mathbf{D} - \mathbf{M}$.

Example 10.35. The graph shown in Fig. 10.4 has the adjacency matrix

$$\mathbf{M} = \begin{bmatrix} 0 & 1 & 1 & 0 & 0 & 0 \\ 1 & 0 & 1 & 0 & 0 & 0 \\ 1 & 1 & 0 & 0 & 0 & 0 \\ 0 & 0 & 0 & 0 & 1 & 1 \\ 0 & 0 & 0 & 1 & 0 & 1 \\ 0 & 0 & 0 & 1 & 1 & 0 \end{bmatrix}.$$

Therefore, it has the Laplacian

$$\mathbf{L} = \begin{bmatrix} 2 & -1 & -1 & 0 & 0 & 0 \\ -1 & 2 & -1 & 0 & 0 & 0 \\ -1 & -1 & 2 & 0 & 0 & 0 \\ 0 & 0 & 0 & 2 & -1 & -1 \\ 0 & 0 & 0 & -1 & 2 & -1 \\ 0 & 0 & 0 & -1 & -1 & 2 \end{bmatrix}.$$

Remark 10.36. Note the row sum of each row in the Laplacian matrix is zero. The Laplacian matrix is also symmetric.

Proposition 10.37. *The Laplacian matrix* $\mathbf{L}$ *of a simple graph* G *is symmetric.* □

Lemma 10.38. *The row sum of the adjacency matrix of a simple graph is the degree of the corresponding vertex.* □

Corollary 10.39. *The row sum for each row of the Laplacian matrix of a simple graph is zero.* □

Theorem 10.40. *Let* $\mathbf{L}$ *be the Laplacian matrix of a simple graph* G. *Suppose* $\mathbf{L} \in \mathbb{R}^{n \times n}$, *then* $\mathbf{1} = \langle 1, 1, \ldots, 1 \rangle \in \mathbb{R}^n$ *is an eigenvector of* $\mathbf{L}$ *with eigenvalue* 0.

Proof. Let

$$\mathbf{L} = \begin{bmatrix} d_{11} & -a_{12} & -a_{13} & \cdots & -a_{1n} \\ -a_{21} & d_{22} & -a_{23} & \cdots & -a_{2n} \\ \vdots & \vdots & \vdots & \ddots & \vdots \\ -a_{n1} & -a_{n2} & -a_{n3} & \cdots & d_{nn} \end{bmatrix}. \tag{10.22}$$

Let $\mathbf{v} = \mathbf{L} \cdot \mathbf{1}$. Recall that $\mathbf{L}_{i\cdot}$ is the ith row of $\mathbf{L}$. The ith component of $\mathbf{v}$ is

$$\begin{aligned} \mathbf{v}_i = \mathbf{L}_{i\cdot} \cdot \mathbf{1} &= \begin{bmatrix} d_{i1} & -a_{i2} & -a_{i3} & \cdots & -a_{in} \end{bmatrix} \begin{bmatrix} 1 \\ 1 \\ 1 \\ \vdots \\ 1 \end{bmatrix} \\ &= d_{i1} - a_{i2} - a_{i3} - \cdots - a_{in} = 0. \end{aligned} \tag{10.23}$$

We have shown that $\mathbf{v}_i = 0$ for $i = 1, \ldots, n$; therefore, $\mathbf{v} = 0$. Thus,

$$\mathbf{L} \cdot \mathbf{1} = \mathbf{0} = 0 \cdot \mathbf{1}.$$

If follows that $\mathbf{1}$ is an eigenvector with eigenvalue 0. This completes the proof. □

Remark 10.41. It is worth noting that 0 can be an eigenvalue, but the zero vector $\mathbf{0}$ cannot be an eigenvector.

Remark 10.42. We know from the principal axis theorem (Theorem A.97) that $\mathbf{L}$ must have n linearly independent (and orthogonal) eigenvectors that form a basis for $\mathbb{R}^n$ since its a real symmetric matrix.

Theorem 10.43. *Let $G = (V, E)$ be a graph with $V = \{v_1, \ldots, v_n\}$ and with Laplacian $\mathbf{L}$. Then, the (algebraic) multiplicity of the eigenvalue 0 is equal to the number of components of G.*

Proof. Assume that G has more than one component; order the components as $H_1, \ldots, H_k$, and suppose that each component has n_i vertices. Then, $n_1 + n_2 + \cdots + n_k = n$. Each component has its own Laplacian matrix $\mathbf{L}_i$ for $i = 1, \ldots, k$, and the Laplacian matrix of G is the block matrix

$$\mathbf{L} = \begin{bmatrix} \mathbf{L}_1 & \mathbf{0} \cdots & \mathbf{0} & \\ \mathbf{0} & \mathbf{L}_2 \cdots & \mathbf{0} & \\ \vdots & \vdots & \ddots & \vdots \\ \mathbf{0} & \mathbf{0} & \cdots & \mathbf{L}_k \end{bmatrix}.$$

Let $\mathbf{1}_i$ be a column vector of 1's that is the eigenvector of $\mathbf{L}_i$ with eigenvalue 0. We can construct an eigenvector of $\mathbf{L}$ as $\mathbf{v}_i = \langle \mathbf{0}, \ldots, \mathbf{1}_i, \mathbf{0}, \ldots, \mathbf{0} \rangle$ with eigenvalue 0. This argument holds for all $\mathbf{L}_1, \ldots, \mathbf{L}_k$. Thus, $\mathbf{L}$ has an eigenvalue of 0 with multiplicity of at least k.

Now, suppose $\mathbf{v}$ is an eigenvector with eigenvalue 0. Then,

$$\mathbf{L}\mathbf{v} = \mathbf{0}.$$

That is, $\mathbf{v}$ is in the null space of $\mathbf{L}$. We have so far proved that the nullity of $\mathbf{L}$ is at least k since each eigenvector $\mathbf{v}_i$ is linearly independent of any other eigenvector $\mathbf{v}_j$ for $i \neq j$. Thus, the basis of the null space of $\mathbf{L}$ contains at least k vectors. On the other hand, it is clear by construction that the rank of the Laplacian matrix $\mathcal{L}_i$ is exactly $n_i - 1$. The structure of $\mathcal{L}$ ensures that the rank of $\mathcal{L}$ is

$$(n_1 - 1) + (n_2 - 1) + \cdots + (n_k - 1) = n - k.$$

From the rank–nullity theorem (see Theorem A.66), we know that the rank of $\mathbf{L}$ plus the nullity of $\mathbf{L}$ must be n. Therefore, the nullity of $\mathbf{L}$ is precisely k. That is, the multiplicity of the eigenvalue 0 is precisely the number of components. This completes the proof. □

Remark 10.44. We state the following fact without proof. Its proof can be found in Ref. [39, Lemma 13.1.1]. It is a consequence of the fact that the Laplacian matrix is positive semi-definite, meaning that for any $\mathbf{v} \in \mathbb{R}^n$, the (scalar) quantity

$$\mathbf{v}^T \mathbf{L} \mathbf{v} \geq 0.$$

Lemma 10.45. *Let G be a graph with Laplacian matrix $\mathcal{L}$. The eigenvalues of $\mathcal{L}$ are all non-negative.* □

Definition 10.46 (Fiedler value/vector). Let G be a simple graph with n vertices with Laplacian $\mathbf{L}$. Suppose $\mathbf{L}$ has eigenvalues $\{\lambda_n, \ldots, \lambda_1\}$ ordered from largest to smallest (i.e., so that $\lambda_n \geq \lambda_{n-1} \geq \cdots \geq \lambda_1$). The second smallest eigenvalue λ_2 is called the *Fiedler value*, and its corresponding eigenvector is called the *Fiedler vector.*

Proposition 10.47. *Let G be a graph with Laplacian matrix $\mathcal{L}$. The Fiedler value $\lambda_2 > 0$ if and only if G is connected.*

Proof. If G is connected, it has one component; therefore, the multiplicity of the 0 eigenvalue is 1. By Lemma 10.45, $\lambda_2 > 0$. On the other hand, suppose that $\lambda_2 > 0$, then, necessarily, $\lambda_1 = 0$ and has a multiplicity of 1. □

Remark 10.48. We state a remarkable fact about the Fiedler vector, whose proof can be found in Ref. [102].

Theorem 10.49. *Let $G = (V, E)$ be a simple graph with $V = \{v_1, \ldots, v_n\}$ and with Laplacian matrix $\mathbf{L}$. If $\mathbf{v}$ is the eigenvector* *corresponding to the Fiedler value λ_2, then the subgraph generated by the set of vertices*

$$V(\mathbf{v}, c) = \{v_i \in V : \mathbf{v}_i \geq c\}$$

is a connected subgraph of G. □

Remark 10.50. In particular, this means that if $c = 0$, then the vertices whose indices correspond to the positive entries in $\mathbf{v}$ allow for a natural bipartition of the vertices of G. This bipartition is called a *spectral clustering* of G or, sometimes, a *Cheeger cut.* This type of clustering can be useful for finding groupings of individuals in social networks.

Example 10.51. Consider the social network shown to the left in Fig. 10.5. If we compute the Fiedler value for this graph, we see that it is $\lambda_2 = 3 - \sqrt{5} > 0$ since the graph is connected. The corresponding Fiedler vector is

$$\mathbf{v} = \begin{bmatrix} \frac{1}{2}\left(-1-\sqrt{5}\right) \\ \frac{1}{2}\left(-1-\sqrt{5}\right) \\ \frac{1}{2}\left(\sqrt{5}-3\right) \\ 1 \\ \frac{1}{2}\left(1+\sqrt{5}\right) \\ 1 \end{bmatrix} \approx \begin{bmatrix} -1.618 \\ -1.618 \\ -0.382 \\ 1. \\ 1.618 \\ 1. \end{bmatrix}.$$

Setting $c = 0$ and assuming that the vertices are in alphabetical order, a natural partition of this social network is

$$V_1 = \{\text{Alice}, \text{Bob}, \text{Cheryl}\} \quad \text{and}$$
$$V_2 = \{\text{David}, \text{Edward}, \text{Finn}\}.$$

That is, we have grouped the vertices together with *negative* entries in the Fiedler vector and grouped the vertices together with *positive* entries in the Fiedler vector. This is illustrated in Fig. 10.5. It is worth noting that if an entry is 0 (i.e., on the border), that vertex can be placed in either partition or placed in a partition of its own. It usually bridges two distinct vertex groups together within the graph structure. For large graphs, this process can be iteratively repeated to produce a spectral clustering.

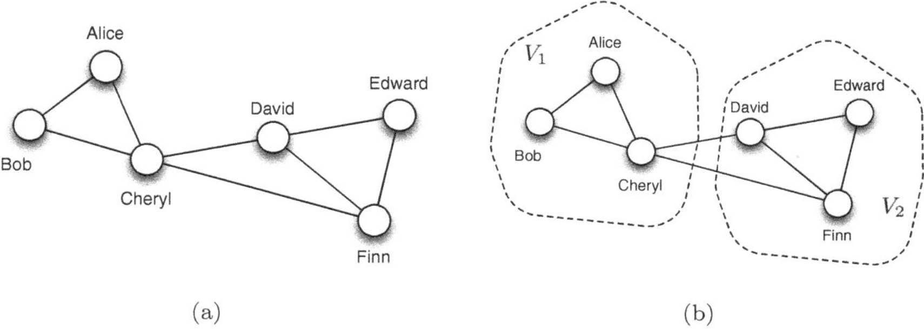

Fig. 10.5 (a) A simple social network and (b) A graph partition using positive and negative entries of the Fiedler vector.

10.5 Chapter Notes

The use of eigenvectors for ranking far predates its by Google in the early 2000s [103]. Dating at least to the 1940s with the work of Seeley [104], it is the social scientists who made the greatest use of these eigenvector-based ranking mechanisms. In 1953, the sociologist Katz [26] developed "Katz centrality," which bears a remarkable similarity to PageRank. Just prior to Katz's work, Wei [105] had used the Perron–Frobenius theorem to put ranking on a firm mathematical footing while specifically focusing on sports team rankings. However, it was Berge [106] who recognized that this approach could be applied to all (directed) graphs. Interestingly, Brin and Page's publications on PageRank [27, 107] lack citation of any of this prior work. We should note that PageRank was extensively covered in the media and the academic literature. See (for example) Refs. [108, 109].

Markov chains were first studied (literally invented) by Russian mathematician Andrey Markov [1]. Together with Kolmogorov, Markov helped establish the foundations of probability theory and stochastic processes. Markov chains have been applied widely in engineering and form the basis of hidden Markov models [110], an early approach to statistical machine learning and a model used extensively in natural language processing.

Miroslav Fiedler, after whom the Fiedler vector was named, was a Czech mathematician whose work in algebraic graph theory [102] paved the way for modern spectral clustering methods. Spectral clustering has been independently developed in both computer science [111] and network science (physics) [9, 112]. However, the graph Laplacian has far more applications than just clustering. The name itself hints at its relation to the Laplacian operator ∇^2, which appears in second-order differential equations (such as the heat equation). In particular, if G is a connected simple graph with Laplacian $\mathbf{L}$, then the graph heat equation is $\dot{\mathbf{u}} = -k\mathbf{L}\mathbf{u}$. It has solutions similar to the continuous heat equation $\partial_t u = k\nabla^2 u$ but on the discrete structure of the graph. This can be useful for solving heat equations with complex boundary conditions. See, for example, Ref. [113] for a specific application. The discrete Laplacian also emerges in the study of consensus on networks, where it frequently drives the dynamics (see, for example, Ref. [114]).

10.6 Exercises

Exercise 10.1
Show that Theorem 10.5 does not hold if there is some other eigenvalue λ of $\mathbf{M}$ so that $|\lambda_0| = |\lambda|$. To do this, consider the path graph with three vertices. Find its adjacency matrix, eigenvalues, and principal eigenvector, and confirm that the theorem does not hold in this case.

Exercise 10.2
Prove Theorem 10.16. [Hint: Use the same inductive argument from the proof of Theorem 9.4.]

Exercise 10.3
Prove Lemma 10.22.

Exercise 10.4
Draw the Markov chain with stochastic matrix

$$\mathbf{M} = \begin{bmatrix} 0 & 1 & 0 \\ 0 & 0 & 1 \\ 1 & 0 & 0 \end{bmatrix}.$$

Show that $\lim_{k\to\infty} \mathbf{M}^k$ does not converge but this Markov chain does have a stationary distribution.

Exercise 10.5
Consider the following Markov chain.

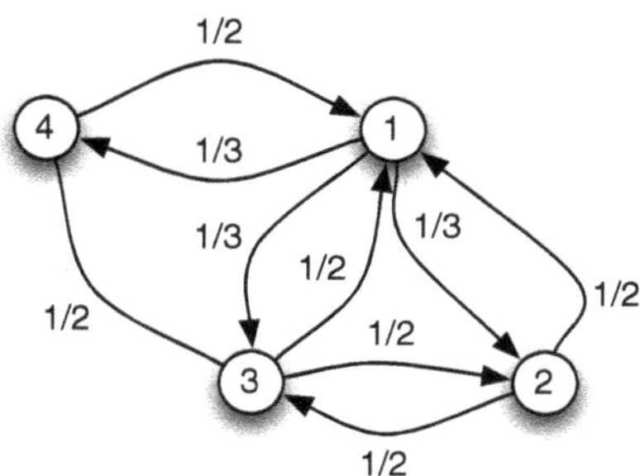

Suppose this is the induced Markov chain from four web pages. Compute the PageRank of these web pages using $d = 0.85$.

Exercise 10.6
Find an expression for $\mathbf{r}^{(2)}$ in terms of $\mathbf{r}^{(0)}$. Explain how the damping factor occurs and how it decreases the chance of taking long walks through the induced Markov chain. Can you generalize your expression for $\mathbf{r}^{(2)}$ to an expression for $\mathbf{r}^{(k)}$ in terms of $\mathbf{r}^{(0)}$?

Exercise 10.7
Prove Proposition 10.37.

Exercise 10.8
Prove Lemma 10.38.

Exercise 10.9
Find a spectral bipartition of the following graph.

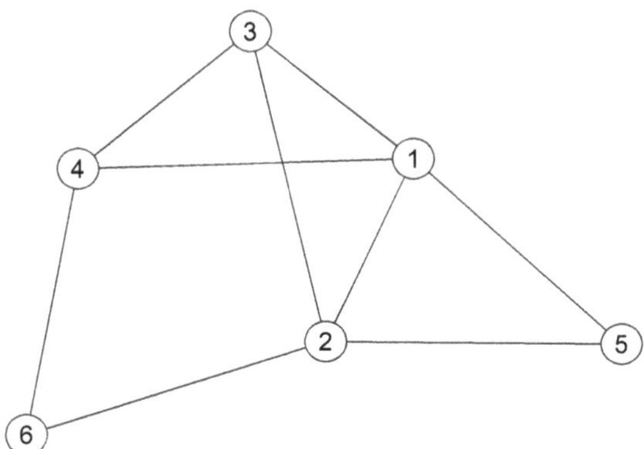

Part 4

Linear Programming and Graph Theory

Chapter 11

A Brief Introduction to Linear Programming

11.1 Introduction and Rationale

Remark 11.1 (Chapter goals). Many graph-theoretic problems can be expressed as linear optimization (sometimes called linear programming) problems. Furthermore, the proofs of some of the most fundamental theorems of graph theory are greatly simplified by the use of a linear optimization formulation.

Even though it seems as if we're going to go far off topic, we use this chapter to introduce linear optimization and its fundamental results. We then use these results to prove the max-flow/min-cut theorem, thereby illustrating the link between the theory of optimization and the theory of graphs.

11.2 Linear Programming: Notation

Definition 11.2 (Linear programming problem). A *linear programming* problem is an optimization problem of the form

$$
\begin{cases}
\max \; z(x_1, \dots, x_n) = c_1 x_1 + \cdots + c_n x_n \\
s.t. \; a_{11} x_1 + \cdots + a_{1n} x_n \leq b_1 \\
\qquad \vdots \\
a_{m1} x_1 + \cdots + a_{mn} x_n \leq b_m \\
h_{11} x_1 + \cdots + h_{n1} x_n = r_1 \\
\qquad \vdots \\
h_{l1} x_1 + \cdots + h_{ln} x_n = r_l.
\end{cases}
\tag{11.1}
$$

Remark 11.3. We can use matrices to write these problems more compactly. Consider the following system of equations:

$$
\begin{cases}
a_{11} x_1 + a_{12} x_2 + \cdots + a_{1n} x_n = b_1 \\
a_{21} x_1 + a_{22} x_2 + \cdots + a_{2n} x_n = b_2 \\
\qquad \vdots \\
a_{m1} x_1 + a_{m2} x_2 + \cdots + a_{mn} x_n = b_m.
\end{cases}
\tag{11.2}
$$

Then, we can write this in matrix notation as

$$\mathbf{Ax} = \mathbf{b}, \tag{11.3}$$

where $\mathbf{A}_{ij} = a_{ij}$ for $i = 1, \dots, m$, $j = 1, \dots, n$, and $\mathbf{x}$ is a column vector in $\mathbb{R}^n$ with entries x_j, $j = 1, \dots, n$, and $\mathbf{b}$ is a column vector in $\mathbb{R}^m$ with entries b_i, $i = 1, \dots, m$. If we replace the equalities in Eq. (11.3) with inequalities, we can also express the systems of inequalities in the form

$$\mathbf{Ax} \leq \mathbf{b}. \tag{11.4}$$

Using this representation, we can write our general linear programming problem using matrix and vector notation. Equation (11.1) becomes

$$
\begin{cases}
\max \; z(\mathbf{x}) = \mathbf{c}^T \mathbf{x} \\
\qquad s.t. \; \mathbf{Ax} \leq \mathbf{b} \\
\qquad \mathbf{Hx} = \mathbf{r}.
\end{cases}
\tag{11.5}
$$

Here, $\mathbf{c}^T$ is the transpose of the column vector $\mathbf{c}$.

Definition 11.4. In Eq. (11.5), if we restrict some of the decision variables (the x_i's) so that they have integer (or discrete) values, then the problem becomes a mixed integer linear programming problem. If all of the variables are restricted to integer values, the problem is an integer programming problem, and if every variable can only take on the values 0 or 1, the program is called a $0-1$ or binary integer programming problem. There are many works on integer programming, of which Ref. [115] is one.

11.3 Intuitive Solutions to Linear Programming Problems

Example 11.5. Consider the problem of a toy company that produces toy planes and toy boats. The toy company can sell its planes for \$10 and its boats for \$8 dollars. It costs \$3 in raw materials to make a plane and \$2 in raw materials to make a boat. A plane requires 3 hours to make and 1 hour to finish, while a boat requires 1 hour to make and 2 hours to finish. The toy company knows it will not sell anymore than 35 planes per week. Furthermore, given the number of workers, the company cannot spend anymore than 160 hours per week finishing toys and 120 hours per week making toys. The company wishes to maximize the profit it makes by choosing how much of each toy to produce.

We can represent the profit maximization problem of the company as a linear programming problem. Let x_1 be the number of planes the company will produce, and let x_2 be the number of boats the company will produce. The profit for each plane is $\$10 - \$3 = \$7$ per plane and the profit for each boat is $\$8 - \$2 = \$6$ per boat. Thus, the total profit the company will make is

$$z(x_1, x_2) = 7x_1 + 6x_2. \tag{11.6}$$

The company can spend no more than 120 hours per week making toys, and since a plane takes 3 hours to make and a boat takes 1 hour to make, we have

$$3x_1 + x_2 \leq 120. \tag{11.7}$$

Likewise, the company can spend no more than 160 hours per week finishing toys, and since it takes 1 hour to finish a plane and 2 hour to finish a boat, we have

$$x_1 + 2x_2 \leq 160. \tag{11.8}$$

Finally, we know that $x_1 \leq 35$ since the company will make no more than 35 planes per week. Thus, the complete linear programming problem is given as

$$\begin{cases} \max & z(x_1, x_2) = 7x_1 + 6x_2 \\ s.t. & 3x_1 + x_2 \leq 120 \\ & x_1 + 2x_2 \leq 160 \\ & x_1 \leq 35 \\ & x_1 \geq 0 \\ & x_2 \geq 0. \end{cases} \tag{11.9}$$

Remark 11.6. To be precise, the linear programming problem in Example 11.5 is not a true linear programming problem because we don't want to manufacture a fractional number of boats or planes and; therefore, x_1 and x_2 must really be drawn from the *integers* and not the real numbers (a requirement for a linear programming problem). However, we ignore this fact and assume that we can indeed manufacture a fractional number of boats and planes.

Remark 11.7. Linear programs (LPs) with two variables can be solved graphically by plotting the feasible region (the values of (x_1, x_2)) that make the inequalities true along with the level curves of the objective function. We show that we can find a point in the feasible region that maximizes the objective function using the level curves of the objective function. We illustrate the method first using the problem from Example 11.5.

Example 11.8 (Continuation of Example 11.5). To solve the linear programming problem from Example 11.5 graphically, begin by drawing the feasible region. That is, plot the inequalities $3x_1 + x_2 \leq 120$, $x_1 + 2x_2 \leq 160$, $x_1 \geq 35$, and $x_1, x_2 \geq 0$. This is shown in the blue shaded region of Fig. 11.1.

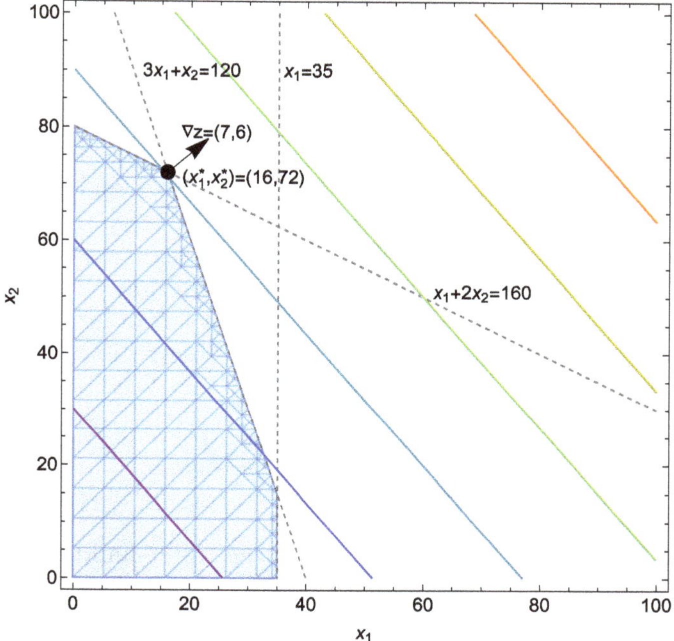

Fig. 11.1 Feasible region and level curves of the objective function: The shaded region in the plot is the feasible region and represents the intersection of the five inequalities constraining the values of x_1 and x_2. The optimal solution is the "last" point in the feasible region that intersects a level set as we move in the direction of increasing profit (the gradient of z).

After plotting the feasible region, the next step is to plot the level curves of the objective function. In our problem, the level sets will have the form

$$7x_1 + 6x_2 = c \qquad \Longrightarrow \qquad x_2 = \frac{-7}{6}x_1 + \frac{c}{6}.$$

This is a set of parallel lines with slope $-7/6$ and intercept $c/6$, where c can be varied as needed. In Fig. 11.1, they are shown in colors ranging from purple to red depending upon the value of c. Larger values of c are more red.

To solve the linear programming problem, follow the level sets along the direction of the gradient of $z = 7x_1 + 6x_2$ (shown as the black arrow) until the last level set (line) intersects the feasible region. The gradient of $7x_1 + 6x_2$ is the vector $(7, 6)$.

When doing this by hand, draw a single line of the form $7x_1 + 6x_2 = c$ and then simply draw parallel lines in the direction of the gradient. At some point, these lines will fail to intersect the feasible region. The last line to intersect the feasible region will do so at a point that maximizes the profit. In this case, the point that maximizes $z(x_1, x_2) = 7x_1 + 6x_2$, subject to the constraints given, is $(x_1^*, x_2^*) = (16, 72)$. This point is the intersection of the two lines $3x_1 + x_2 = 120$ and $x_1 + 2x_2 = 160$. Note that the point of optimality $(x_1^*, x_2^*) = (16, 72)$ is at a corner of the feasible region. In this case, the constraints

$$\begin{aligned} 3x_1 + x_2 &\leq 120, \\ x_1 + 2x_2 &\leq 160 \end{aligned}$$

are both *binding* (equal to their respective right-hand sides), while the other constraints are nonbinding. In general, we see that when an optimal solution to a linear programming problem exists, it will always be at the intersection of several binding constraints; that is, it will occur at a corner of a higher-dimensional polyhedron.

Remark 11.9. It can sometimes happen that a linear programming problem has an infinite number of alternative optimal solutions. We illustrate this in the following example.

Example 11.10. Suppose the toy maker in Example 11.5 finds that it can sell planes for a profit of \$18 each instead of \$7 each. The new linear programming problem becomes

$$\begin{cases} \max \ z(x_1, x_2) = 18x_1 + 6x_2 \\ s.t. \ 3x_1 + x_2 \leq 120 \\ \quad x_1 + 2x_2 \leq 160 \\ \quad x_1 \leq 35 \\ \quad x_1 \geq 0 \\ \quad x_2 \geq 0. \end{cases} \tag{11.10}$$

Applying the graphical method for finding optimal solutions to linear programming problems yields the plot shown in Fig. 11.2. The level

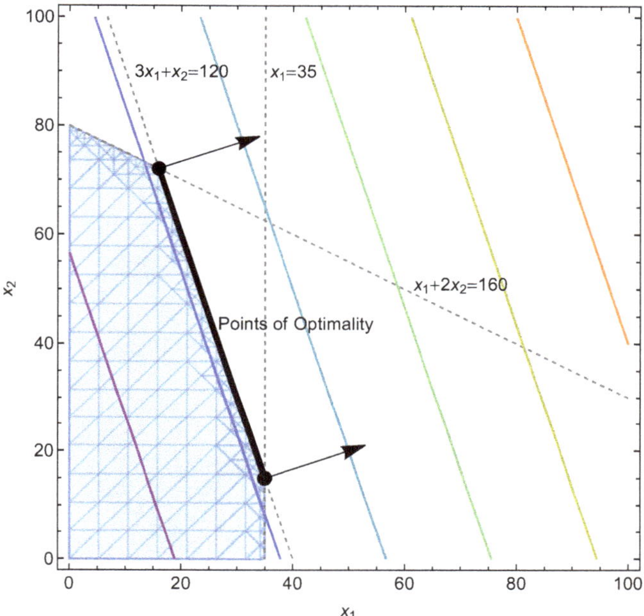

Fig. 11.2 An example of infinitely many alternative optimal solutions in a linear programming problem. The level curves for $z(x_1, x_2) = 18x_1 + 6x_2$ are *parallel* to one face of the polygon boundary of the feasible region. Moreover, this side contains the points of the greatest value for $z(x_1, x_2)$ inside the feasible region. Any combination of (x_1, x_2) on the line $3x_1 + x_2 = 120$ for $x_1 \in [16, 35]$ will provide the largest possible value that $z(x_1, x_2)$ can take in the feasible region S.

curves for the function $z(x_1, x_2) = 18x_1 + 6x_2$ are *parallel* to one face (edge) of the polygonal boundary of the feasible region. Hence, as we move further up and to the right in the direction of the gradient (corresponding to larger and larger values of $z(x_1, x_2)$), we see that there is not *one* point on the boundary of the feasible region that intersects that level set with the greatest value, but, instead, a side of the polygonal boundary described by the line $3x_1 + x_2 = 120$, where $x_1 \in [16, 35]$. Let

$$S = \{(x_1, x_2 | 3x_1 + x_2 \leq 120,\ x_1 + 2x_2 \leq 160,\ x_1 \leq 35,\ x_1, x_2 \geq 0\}.$$

That is, S is the feasible region of the problem. Then, for any value of $x_1^* \in [16, 35]$ and any value x_2^* so that $3x_1^* + x_2^* = 120$, we have $z(x_1^*, x_2^*) \geq z(x_1, x_2)$ for all $(x_1, x_2) \in S$. Since there are infinitely many values that x_1 and x_2 may take on, we see this problem has an infinite number of alternative optimal solutions.

11.4 Some Basic Facts about Linear Programming Problems

Definition 11.11 (Canonical form). A maximization linear programming problem is in *canonical form* if it is written as

$$\begin{cases} \max \quad z(\mathbf{x}) = \mathbf{c}^T\mathbf{x} \\ \qquad\qquad s.t.\ \mathbf{A}\mathbf{x} \leq \mathbf{b} \\ \qquad\qquad\qquad \mathbf{x} \geq \mathbf{0}. \end{cases} \tag{11.11}$$

A minimization linear programming problem is in *canonical form* if it is written as

$$\begin{cases} \min \quad z(\mathbf{x}) = \mathbf{c}^T\mathbf{x} \\ \qquad\qquad s.t.\ \mathbf{A}\mathbf{x} \geq \mathbf{b} \\ \qquad\qquad\qquad \mathbf{x} \geq \mathbf{0}. \end{cases} \tag{11.12}$$

Definition 11.12 (Standard form). A linear programming problem is in *standard form* if it is written as

$$\begin{cases} \max \quad z(\mathbf{x}) = \mathbf{c}^T\mathbf{x} \\ \qquad\qquad s.t.\ \mathbf{A}\mathbf{x} = \mathbf{b} \\ \qquad\qquad\qquad \mathbf{x} \geq \mathbf{0}. \end{cases} \tag{11.13}$$

Remark 11.13. The following theorem is outside the scope of the course, but it is useful to know.

Theorem 11.14. *Every linear programming problem in canonical form can be put into standard form.* □

Remark 11.15. To illustrate Theorem 11.14, we note that it is relatively easy to convert any inequality constraint into an equality constraint. Consider the inequality constraint

$$a_{i1}x_1 + a_{i2}x_2 + \cdots + a_{in}x_n \leq b_i. \tag{11.14}$$

We can add a new *slack variable* s_i to this constraint to obtain

$$a_{i1}x_1 + a_{i2}x_2 + \cdots + a_{in}x_n + s_i = b_i.$$

Obviously, this slack variable $s_i \geq 0$. The slack variable then becomes just another variable whose value we must discover as we solve the linear program for which Eq. (11.14) is a constraint.

We can deal with constraints of the form

$$a_{i1}x_1 + a_{i2}x_2 + \cdots + a_{in}x_n \geq b_i \tag{11.15}$$

in a similar way. In this case, we subtract a surplus variable s_i to obtain

$$a_{i1}x_1 + a_{i2}x_2 + \cdots + a_{in}x_n - s_i = b_i.$$

Again, we must have $s_i \geq 0$.

Example 11.16. Consider the linear programming problem

$$\begin{cases} \max & z(x_1, x_2) = 2x_1 - x_2 \\ s.t. & x_1 - x_2 \leq 1 \\ & 2x_1 + x_2 \geq 6 \\ & x_1, x_2 \geq 0. \end{cases}$$

This linear programming problem can be put into standard form by using both a slack and a surplus variable. We obtain

$$\begin{cases} \max & z(x_1, x_2) = 2x_1 - x_2 \\ s.t. & x_1 - x_2 + s_1 = 1 \\ & 2x_1 + x_2 - s_2 = 6 \\ & x_1, x_2, s_1, s_2 \geq 0. \end{cases}$$

Remark 11.17. We assume, when dealing with linear programming problems in standard or canonical form, that the matrix $\mathbf{A}$ has full row rank, and if not, we adjust it so this is true. The following theorem fully characterizes the solutions to linear programs. The proof can be found in Ref. [29].

Theorem 11.18. *Consider any linear programming problem*

$$P \begin{cases} \max & z(\mathbf{x}) = \mathbf{c}^T\mathbf{x} \\ & s.t. \ \mathbf{A}\mathbf{x} \leq \mathbf{b} \\ & \mathbf{x} \geq \mathbf{0}. \end{cases}$$

Then, there are exactly four possibilities:

(1) *There is a unique solution to problem P denoted by $\mathbf{x}^*$.*
(2) *There are an infinite number of alternative optimal solutions to P.*
(3) *There is no solution to P because there is no $\mathbf{x}$ that satisfies $\mathbf{Ax} \leq \mathbf{b}$ and $\mathbf{x} \geq \mathbf{0}$.*
(4) *There is no solution to P because the problem is unbounded. That is, for any $\mathbf{x}$ such that $\mathbf{Ax} = \mathbf{b}$, there is another $\mathbf{x}' \neq \mathbf{x}$ so that $\mathbf{Ax}' = \mathbf{b}$ and $\mathbf{c}^T\mathbf{x} < \mathbf{c}^T\mathbf{x}'$.* □

11.5 Solving Linear Programming Problems with a Computer

Remark 11.19. There are a few ways to solve linear programming problems. The most common approach is called the *simplex algorithm*. The simplex algorithm is outside the scope of this book. However, those interested can see Ref. [29].

Example 11.20. We illustrate how to solve a linear programming problem using Mathematica™ because it is particularly easy to do so. Suppose I wish to design a diet consisting of Ramen noodles and ice cream. I'm interested in spending as little money as possible, but I want to ensure that I eat at least 1200 calories per day and that I get at least 20 grams of protein per day. Assume that each serving of Ramen costs \$1 and contains 100 calories and 2 grams of protein. Assume that each serving of ice cream costs \$1.50 and contains 200 calories and 3 grams of protein.

We can construct a linear programming problem out of this scenario. Let x_1 be the amount of Ramen I consume and x_2 be the amount of ice cream I consume. Our objective function is the cost

$$x_1 + 1.5x_2. \tag{11.16}$$

The constraints describe our protein requirements as

$$2x_1 + 3x_2 \geq 20 \tag{11.17}$$

and our calorie requirements (expressed in terms of hundreds of calories)

$$x_1 + 2x_2 \geq 12. \tag{11.18}$$

This leads to the following linear programming problem

$$\begin{cases} \min \ x_1 + 1.5x_2 \\ s.t. \ \ 2x_1 + 3x_2 \geq 20 \\ \qquad x_1 + 2x_2 \geq 12 \\ \qquad x_1, x_2 \geq 0. \end{cases} \tag{11.19}$$

In Mathematica™, comments are written as `(*Comment*)`. The Mathematica™ code to solve this problem is shown as follows.

```
FindMinimum[
 {
      x1 + 1.5*x2,      (*Objective*)
    2*x1 + 3*x2 >= 20, (*Constraint 1*)
      x1 + 2*x2 >= 12, (*Constraint 2*)
      x1 >= 0, x2 >= 0 (*Non-negativity Constraints*)
  },
  {x1, x2} (*Variables*)
 ]
```

Note that it is relatively easy to interpret this code because Mathematica™ is a symbolic language. Other solvers (such as Python and MATLAB™) can also be used. The solution returned by the solver is $x_1 = x_2 = 4$ with a cost of \$10. It turns out there are an infinite number of alternative optimal solutions to this problem, which can be demonstrated through a diagram.

11.6 Karush–Kuhn–Tucker Conditions

Remark 11.21. The single most important thing to learn about linear programming (or optimization in general) is the Karush–Kuhn–Tucker (KKT) theorem giving optimality conditions. These conditions provide necessary and sufficient conditions for a point

$\mathbf{x} \in \mathbb{R}^n$ to be an optimal solution to a linear programming problem. We state the Karush–Kuhn–Tucker theorem but do not prove it. A proof can be found in Ref. [29].

Theorem 11.22. *Consider the linear programming problem*

$$P\begin{cases} \max & \mathbf{cx} \\ s.t. & \mathbf{Ax} \leq \mathbf{b} \\ & \mathbf{x} \geq \mathbf{0}, \end{cases} \tag{11.20}$$

with $\mathbf{A} \in \mathbb{R}^{m \times n}$, $\mathbf{b} \in \mathbb{R}^m$ *and (row vector)* $\mathbf{c} \in \mathbb{R}^n$. *Then,* $\mathbf{x}^* \in \mathbb{R}^n$ *is an optimal solution to Problem* P *if and only if there exists (row) vectors* $\mathbf{w}^* \in \mathbb{R}^m$ *and* $\mathbf{v}^* \in \mathbb{R}^n$ *and a slack variable vector* $\mathbf{s}^* \in \mathbf{R}^m$ *so that*

$$\textit{primal feasibility}\begin{cases} \mathbf{Ax}^* + \mathbf{s}^* = \mathbf{b} \\ \mathbf{x}^* \geq \mathbf{0} \end{cases} \tag{11.21}$$

$$\textit{dual feasibility}\begin{cases} \mathbf{w}^*\mathbf{A} - \mathbf{v}^* = \mathbf{c} \\ \mathbf{w}^* \geq \mathbf{0} \\ \mathbf{v}^* \geq \mathbf{0} \end{cases} \tag{11.22}$$

$$\textit{complementary slackness}\begin{cases} \mathbf{w}^*\left(\mathbf{Ax}^* - \mathbf{b}\right) = 0 \\ \mathbf{v}^*\mathbf{x}^* = 0. \end{cases} \tag{11.23}$$

□

Remark 11.23. The vectors $\mathbf{w}^*$ and $\mathbf{v}^*$ are sometimes called *dual variables* for reasons that will be clear shortly. They are also sometimes called *Lagrange multipliers*. You may have encountered Lagrange multipliers in vector calculus. These are the same kind of variables, except applied to linear optimization problems. There is one element in the dual variable vector $\mathbf{w}^*$ for each constraint of the form $\mathbf{Ax} \leq \mathbf{b}$ and one element in the dual variable vector $\mathbf{v}^*$ for each constraint of the form $\mathbf{x} \geq \mathbf{0}$.

Example 11.24. Consider the toy maker problem (Eq. (11.9)) with dual variables (Lagrange multipliers) listed next to their corresponding constraints:

$$\begin{cases} \max \ z(x_1, x_2) = 7x_1 + 6x_2 & \textbf{Dual Variable} \\ s.t. \ 3x_1 + x_2 \leq 120 & (w_1) \\ \quad x_1 + 2x_2 \leq 160 & (w_1) \\ \quad x_1 \leq 35 & (w_3) \\ \quad x_1 \geq 0 & (v_1) \\ \quad x_2 \geq 0 & (v_2). \end{cases}$$

In this problem, we have

$$\mathbf{A} = \begin{bmatrix} 3 & 1 \\ 1 & 2 \\ 1 & 0 \end{bmatrix} \quad \mathbf{b} = \begin{bmatrix} 120 \\ 160 \\ 35 \end{bmatrix} \quad \mathbf{c} = \begin{bmatrix} 7 & 6 \end{bmatrix}.$$

Then, the KKT conditions can be written as

$$\text{primal feasibility} \begin{cases} \begin{bmatrix} 3 & 1 \\ 1 & 2 \\ 1 & 0 \end{bmatrix} \begin{bmatrix} x_1 \\ x_2 \end{bmatrix} \leq \begin{bmatrix} 120 \\ 160 \\ 35 \end{bmatrix} \\ \begin{bmatrix} x_1 \\ x_2 \end{bmatrix} \geq \begin{bmatrix} 0 \\ 0 \end{bmatrix} \end{cases}$$

$$\text{dual feasibility} \begin{cases} \begin{bmatrix} w_1 & w_2 & w_3 \end{bmatrix} \begin{bmatrix} 3 & 1 \\ 1 & 2 \\ 1 & 0 \end{bmatrix} - \begin{bmatrix} v_1 & v_2 \end{bmatrix} = \begin{bmatrix} 7 & 6 \end{bmatrix} \\ \begin{bmatrix} w_1 & w_2 & w_3 \end{bmatrix} \geq \begin{bmatrix} 0 & 0 & 0 \end{bmatrix} \\ \begin{bmatrix} v_1 & v_2 \end{bmatrix} \geq \begin{bmatrix} 0 & 0 \end{bmatrix} \end{cases}$$

$$\text{complementary slackness} \begin{cases} \begin{bmatrix} w_1 & w_2 & w_3 \end{bmatrix} \left(\begin{bmatrix} 3 & 1 \\ 1 & 2 \\ 1 & 0 \end{bmatrix} \begin{bmatrix} x_1 \\ x_2 \end{bmatrix} - \begin{bmatrix} 120 \\ 160 \\ 35 \end{bmatrix} \right) = 0 \\ \begin{bmatrix} v_1 & v_2 \end{bmatrix} \begin{bmatrix} x_1 & x_2 \end{bmatrix} = 0. \end{cases}$$

Note that we are suppressing the slack variables $\mathbf{s}$ in the primal feasibility expression for brevity. Recall that at optimality, we had $x_1 = 16$ and $x_2 = 72$. The binding constraints in this case were

$$3x_1 + x_2 \leq 120 \quad \text{and}$$
$$x_1 + 2x_2 \leq 160.$$

To see this, note that if $3(16)+72 = 120$ and $16+2(72) = 160$. Then, we should be able to express $\mathbf{c} = [7 \quad 6]$ (the vector of coefficients of the objective function) as a positive combination of the gradients of the binding constraints

$$\nabla(7x_1 + 6x_2) = \begin{bmatrix} 7 & 6 \end{bmatrix}$$
$$\nabla(3x_1 + x_2) = \begin{bmatrix} 3 & 1 \end{bmatrix}$$
$$\nabla(x_1 + 2x_2) = \begin{bmatrix} 1 & 2 \end{bmatrix}.$$

This is what dual feasibility asserts to be true. That is, we wish to solve the linear equation

$$\begin{bmatrix} w_1 & w_2 \end{bmatrix} \begin{bmatrix} 3 & 1 \\ 1 & 2 \end{bmatrix} = \begin{bmatrix} 7 & 6 \end{bmatrix}. \tag{11.24}$$

The result is the system of equations

$$3w_1 + w_2 = 7,$$
$$w_1 + 2w_2 = 6.$$

A solution to this system is $w_1 = \frac{8}{5}$ and $w_2 = \frac{11}{5}$. This fact is illustrated in Fig. 11.3.

Figure 11.3 shows that the gradient lies in the cone formed by the gradients of the binding constraints at the optimal point for the toy maker problem. This is generally true. At a point of optimality, the gradient will lie inside the cone generated by the gradients of the binding constraints. Since $x_1, x_2 > 0$, we must have $v_1 = v_2 = 0$. Moreover, since $x_1 < 35$, we know that $x_1 \leq 35$ is not a binding constraint and thus its dual variable w_3 is also zero. This leads to the conclusion that

$$\begin{bmatrix} x_1^* \\ x_2^* \end{bmatrix} = \begin{bmatrix} 16 \\ 72 \end{bmatrix} \quad \begin{bmatrix} w_1^* & w_2^* & w_3^* \end{bmatrix} = \begin{bmatrix} 8/5 & 11/5 & 0 \end{bmatrix} \quad \begin{bmatrix} v_1^* & v_2^* \end{bmatrix} = \begin{bmatrix} 0 & 0 \end{bmatrix},$$

and the KKT conditions are satisfied.

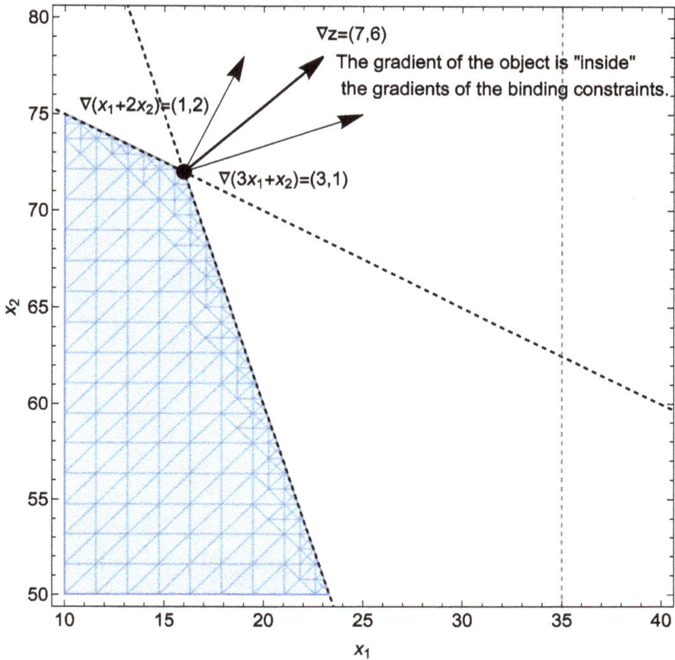

Fig. 11.3 The gradient cone: At optimality, the cost vector **c** is obtuse with respect to the directions formed by the binding constraints. It is also contained inside the cone of the gradients of the binding constraints.

11.7 Duality

Remark 11.25. In this section, we show that to each linear programming problem (the primal problem), we may associate another linear programming problem (the dual linear programming problem). These two problems are closely related to each other, and an analysis of the dual problem can provide deep insight into the primal problem.

Definition 11.26 (Dual program). Consider the linear programming problem

$$P\begin{cases} \max \ \mathbf{c}^T\mathbf{x} \\ s.t. \ \mathbf{A}\mathbf{x} \leq \mathbf{b} \\ \quad\ \ \mathbf{x} \geq \mathbf{0}. \end{cases} \tag{11.25}$$

Then, the dual problem for Problem P is

$$D\begin{cases}\min & \mathbf{wb}\\ s.t. & \mathbf{wA} \geq \mathbf{c}\\ & \mathbf{w} \geq \mathbf{0}.\end{cases} \tag{11.26}$$

Remark 11.27. Let $\mathbf{v}$ be a vector of *surplus* variables. Then, we can transform Problem D into standard form as

$$D_S\begin{cases}\min & \mathbf{wb}\\ s.t. & \mathbf{wA} - \mathbf{v} = \mathbf{c}\\ & \mathbf{w} \geq \mathbf{0}\\ & \mathbf{v} \geq \mathbf{0}\end{cases} \tag{11.27}$$

Thus, we already see an intimate relationship between duality and the KKT conditions. The feasible region of the dual problem (in standard form) is precisely the dual feasibility constraints of the KKT conditions for the primal problem.

In this formulation, we see that we have assigned a dual variable w_i $(i = 1, \dots, m)$ to each constraint in the system of equations $\mathbf{Ax} \leq \mathbf{b}$ of the primal problem. Likewise, dual variables $\mathbf{v}$ can be thought of as corresponding to the constraints in $\mathbf{x} \geq \mathbf{0}$.

Remark 11.28. The proof of the following lemma can be found in Ref. [29].

Lemma 11.29. *The dual of the dual problem is the primal problem.* □

Remark 11.30. Lemma 11.29 shows that the notions of dual and primal can be exchanged and that it is simply a matter of perspective which problem is the dual problem and which is the primal problem. Likewise, by transforming problems into canonical forms, we can develop dual problems for any linear programming problem.

The process of developing these formulations can be exceptionally tedious as it requires enumeration of all the possible combinations of various linear and variable constraints. The following figure summarizes the process of converting an arbitrary primal problem into its dual.

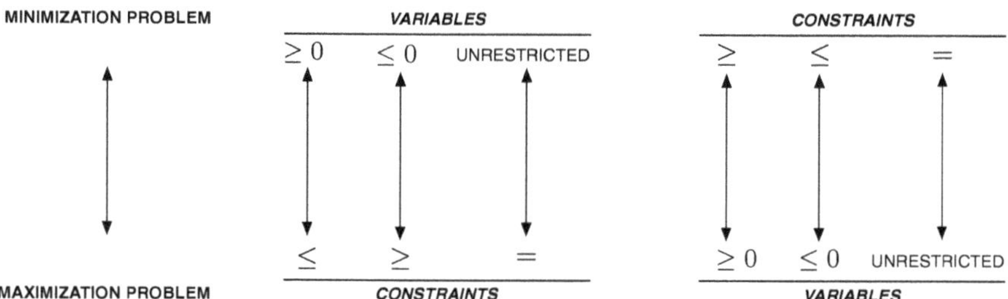

Fig. 11.4 Table of dual conversions: To create a dual problem, assign a dual variable to each constraint of the form $\mathbf{A}\mathbf{x}\square\mathbf{b}$, where $\square$ represents a binary relation. Then, use the chart to determine the appropriate sign of the inequality in the dual problem, as well as the nature of the dual variables. The signs of the primal variables are then used to determine the structure of the dual constraints.

Example 11.31. Consider the problem of finding the dual problem for the toy maker problem (Example 11.5) in standard form. The primal problem is

$$\begin{aligned}
\max\ & 7x_1 + 6x_2 \\
s.t.\ & 3x_1 + x_2 + s_1 = 120 \qquad (w_1) \\
& x_1 + 2x_2 + s_2 = 160 \qquad (w_2) \\
& x_1 + s_3 = 35 \qquad (w_3) \\
& x_1, x_2, s_1, s_2, s_3 \geq 0.
\end{aligned}$$

Here, we have placed dual variable names (w_1, w_2, and w_3) next to the constraints to which they correspond.

The primal problem variables in this case are all positive. So, using Fig. 11.4, we know that the constraints of the dual problem will be greater-than-or-equal-to constraints. Likewise, we know that the dual variables will be unrestricted in sign since the primal problem constraints are all equality constraints.

The coefficient matrix is

$$\mathbf{A} = \begin{bmatrix} 3 & 1 & 1 & 0 & 0 \\ 1 & 2 & 0 & 1 & 0 \\ 1 & 0 & 0 & 0 & 1 \end{bmatrix}.$$

We also have

$$\mathbf{c} = \begin{bmatrix} 7 & 6 & 0 & 0 & 0 \end{bmatrix} \quad \text{and}$$

$$\mathbf{b} = \begin{bmatrix} 120 \\ 160 \\ 35 \end{bmatrix}.$$

Since $\mathbf{w} = [w_1 \ \ w_2 \ \ w_3]$, we know that $\mathbf{wA}$ will be

$$\mathbf{wA} = \begin{bmatrix} 3w_1 + w_2 + w_3 & w_1 + 2w_2 & w_1 & w_2 & w_3 \end{bmatrix}.$$

This vector will be related to $\mathbf{c}$ in the constraints of the dual problem. Remember that in this case, all variables in the primal problem are greater-than-or-equal-to zero. Thus, we see that the constraints of the dual problem are

$$\begin{aligned} 3w_1 + w_2 + w_3 &\geq 7 \\ w_1 + 2w_2 &\geq 6 \\ w_1 &\geq 0 \\ w_2 &\geq 0 \\ w_3 &\geq 0. \end{aligned}$$

We also have the redundant set of constraints that tells us that $\mathbf{w}$ is unrestricted because the primal problem had equality constraints. This *will always* happen in cases when you've introduced slack variables into a problem to put it in standard form. This should be clear from the definition of the dual problem for a maximization problem in canonical form.

Thus, the whole dual problem becomes

$$\begin{aligned} \min \quad & 120w_1 + 160w_2 + 35w_3 \\ s.t. \quad & 3w_1 + w_2 + w_3 \geq 7 \\ & w_1 + 2w_2 \geq 6 \\ & w_1 \geq 0 \\ & w_2 \geq 0 \\ & w_3 \geq 0 \\ & \mathbf{w} \quad \text{unrestricted.} \end{aligned} \tag{11.28}$$

Again, note that, in reality, the constraints we derived from the $\mathbf{wA} \geq \mathbf{c}$ part of the dual problem make the constraints "$\mathbf{w}$ unrestricted" redundant, for in fact $\mathbf{w} \geq \mathbf{0}$, just as we would expect it to be if we'd found the dual of the toy maker problem given in canonical form.

Theorem 11.32 (Strong duality theorem). *Consider Problem P and Problem D. Then,*

(weak duality) $\mathbf{cx}^* \leq \mathbf{w}^*\mathbf{b}$, *thus every feasible solution to the primal problem provides a lower bound for the dual and every feasible solution to the dual problem provides an upper bound to the primal problem.*

Furthermore, exactly one of the following statements is true:

(1) *Both Problem P and Problem D possess optimal solutions $\mathbf{x}^*$ and $\mathbf{w}^*$, respectively, and $\mathbf{cx}^* = \mathbf{w}^*\mathbf{b}$.*
(2) *Problem P is unbounded and Problem D is infeasible.*
(3) *Problem D is unbounded and Problem P is infeasible.*
(4) *Both problems are infeasible.* □

Remark 11.33. This final theorem illustrates the true nature of duality. Two linear programming problems are dual if they share KKT conditions, i.e., if they share conditions for optimality. We conclude that the KKT conditions are fundamental. The optimization problems are simply expressions of these systems of equations and inequalities.

Theorem 11.34. *Problem D has an optimal solution $\mathbf{w}^* \in \mathbb{R}^m$ if and only if there exists vectors $\mathbf{x}^* \in \mathbb{R}^n$ and $\mathbf{s}^* \in \mathbb{R}^m$ and a vector of surplus variables $\mathbf{v}^* \in \mathbb{R}^n$ such that:*

$$\textit{primal feasibility}\begin{cases} \mathbf{w}^*\mathbf{A} \geq \mathbf{c} \\ \mathbf{w}^* \geq \mathbf{0} \end{cases} \tag{11.29}$$

$$\textit{dual feasibility}\begin{cases} \mathbf{Ax}^* + \mathbf{s}^* = \mathbf{b} \\ \mathbf{x}^* \geq \mathbf{0} \\ \mathbf{s}^* \geq \mathbf{0} \end{cases} \tag{11.30}$$

$$\text{complementary slackness} \begin{cases} (\mathbf{w}^*\mathbf{A} - \mathbf{c})\,\mathbf{x}^* = 0 \\ \mathbf{w}^*\mathbf{s}^* = 0. \end{cases} \tag{11.31}$$

Furthermore, these KKT conditions are equivalent to the KKT conditions for the primal problem. □

11.8 Chapter Notes

The problem of solving a system of linear inequalities was studied as early as in the 1800s, and a solution was given by Fourier. This method is now called the Fourier–Motzkin method [116] and is not usually covered in modern treatments of linear programming (except in its historical context). Foundational contributions to linear programming were made by George B. Dantzig, whose solution of two open problems in statistics (mistaken for homework while as a student) [117–119] may have formed the basis for part of the script of *Good Will Hunting*. Dantzig developed the simplex algorithm for solving arbitrary linear programming problems and is considered the father of modern algorithmic optimization. He first proved the duality results discussed in this chapter, but it was Von Neumann who conjectured them on meeting with Dantzig to discuss linear programming [120]. Dantzig's work (at first derided by colleagues for being linear in a nonlinear world) was defended by Von Neumann. Interestingly, linear programming problems are often at the core of many problems in optimization [118, 119]. His work on flows in networks with Fulkerson will be covered in the following chapter.

Amazingly, though the simplex algorithm is, at the worst case, exponential in running time, solving a linear programming problem can be accomplished in polynomial time [121]. This relatively recent result was improved and made practical by Karmarker's interior-point method [122], which has spawned an entirely new field in computational optimization. Interestingly, the linear programs that arise in the study of flow on graphs are known to be solvable in polynomial time (see Chapter 6). Even more amazingly, in 2004, Spielman and Teng proved that the average running time for the simplex algorithm is, in fact, polynomial [123], helping to explain decades of observations on the general efficiency of that algorithm despite it being an exponential algorithm in the worst case.

11.9 Exercises

Exercise 11.1
Show that a minimization linear programming problem in canonical form can be rephrased as a maximization linear programming problem in canonical form. [Hint: Multiply the objective and constraints -1. Define new matrices.]

Exercise 11.2
Consider the problem:

$$\begin{aligned} \max\ & x_1 + x_2 \\ s.t.\ & 2x_1 + x_2 \leq 4 \\ & x_1 + 2x_2 \leq 6 \\ & x_1, x_2 \geq 0. \end{aligned}$$

Write the KKT conditions for an optimal point for this problem. (You will have a vector $\mathbf{w} = [w_1 \quad w_2]$ and a vector $\mathbf{v} = [v_1 \quad v_2]$).

Draw the feasible region of the problem, and use MATLAB™ to solve the problem. At the point of optimality, identify the binding constraints and draw their gradients. Show that the KKT conditions hold. (Specifically, find $\mathbf{w}$ and $\mathbf{v}$.)

Exercise 11.3
Find the KKT conditions for the problem

$$\begin{cases} \min\ \mathbf{cx} \\ s.t.\ \mathbf{Ax} \geq \mathbf{b} \\ \qquad \mathbf{x} \geq \mathbf{0}. \end{cases} \tag{11.32}$$

[Hint: Remember that every minimization problem can be converted to a maximization problem by multiplying the objective function by -1, and the constraints $\mathbf{Ax} \geq \mathbf{b}$ are equivalent to the constraints $-\mathbf{Ax} \leq -\mathbf{b}$.]

Exercise 11.4
Identify the dual problem for

$$\begin{aligned} \max \ & x_1 + x_2 \\ s.t. \ & 2x_1 + x_2 \geq 4 \\ & x_1 + 2x_2 \leq 6 \\ & x_1, x_2 \geq 0. \end{aligned}$$

Exercise 11.5
Use the table or the definition of duality to determine the dual for the problem

$$\begin{cases} \min \ \mathbf{cx} \\ s.t. \ \mathbf{Ax} \geq \mathbf{b} \\ \quad\ \ \mathbf{x} \geq \mathbf{0}. \end{cases} \tag{11.33}$$

Chapter 12

Max Flow/Min Cut with Linear Programming

Remark 12.1 (Chapter goals). The goal of this chapter is to discuss the maximum flow problem using a linear programming approach. We prove again the max-flow/min-cut theorem using linear programming formalisms and show that the duality of edge flows and vertex cuts follows from linear programming duality.

Remark 12.2. In Chapter 6, we proved the max-flow/min-cut theorem using a standard argument without appealing to the fact that it is a linear programming problem. In this chapter, we reconsider that problem and show how to phrase the maximum flow problem as a linear programming problem and also study its dual.

12.1 The Maximum Flow Problem as a Linear Program

Remark 12.3. Recall first the flow conservation constraint, which will now be viewed in the context of a linear programming problem.

Definition 12.4 (Flow conservation constraint). Let $G = (V, E)$ be a digraph with no self-loops and suppose $V = \{v_1, \ldots, v_m\}$ and $E = \{e_1, \ldots, e_n\}$. Let $I(i)$ be the set of edges with destination vertex v_i and $O(i)$ be the set of edges with source v_i. Then, the flow conservation constraint associated with vertex v_i is

$$\sum_{k \in O(i)} x_k - \sum_{k \in I(i)} x_k = b_i \quad \forall i. \tag{12.1}$$

Here, x_k is the flow on edge e_k and b_i is the vertex supply (or demand) at vertex v_i.

Remark 12.5. Remember that Eq. (12.1) states that the total flow out of vertex v_i minus the total flow into v_i must be equal to the total flow produced at v_i. Put more simply, excess flow is neither created nor destroyed.

Proposition 12.6. *Let $G = (V, E)$ be a digraph with no self-loops and suppose $V = \{v_1, \ldots, v_n\}$. Let $\mathbf{A}$ be the incidence matrix of G (see Definition* 9.10*). Then, Eq.* (12.1) *can be written as*

$$\mathbf{A}_{i \cdot}\mathbf{x} = b_i, \tag{12.2}$$

where $\mathbf{x}$ is a vector of variables of the form x_k taken in the order the edges are represented in $\mathbf{A}$.

Proof. From Definition 9.10, we know that

$$\mathbf{A}_{ik} = \begin{cases} 0 & \text{if } v_i \text{ is not in } e_k, \\ 1 & \text{if } v_i \text{ is the source of } e_k, \\ -1 & \text{if } v_i \text{ is the destination of } e_k. \end{cases} \tag{12.3}$$

The equivalence between Eq. (12.2) and Eq. (12.1) follows at once from this fact. □

Remark 12.7. Recall that the standard basis vector $\mathbf{e}_i \in \mathbb{R}^{m \times 1}$ has a 1 at position i and 0 everywhere else.

Definition 12.8 (Maximum flow problem). Let $G = (V, E)$ be a digraph with no self-loops and suppose that $V = \{v_1, \ldots, v_m\}$. Without loss of generality, suppose that there is no edge connecting v_m to v_1. The *maximum flow problem* for G is the linear programming problem

$$\begin{cases} \max & f \\ s.t. & (\mathbf{e}_m - \mathbf{e}_1) f + \mathbf{Ax} = \mathbf{0} \\ & \mathbf{x} \leq \mathbf{u} \\ & \mathbf{x} \geq \mathbf{0} \\ & f \text{ unrestricted.} \end{cases} \tag{12.4}$$

Here, $\mathbf{u}$ is a vector of edge flow capacity values.

Remark 12.9. The constraints $(\mathbf{e}_m - \mathbf{e}_1) f + \mathbf{Ax} = \mathbf{0}$ are flow conservation constraints when we assume that there is an (imaginary) flow backward from v_m to v_1 along an edge (v_m, v_1) and that no flow is produced in the graph. That is, we assume that all flow is circulating within the graph. The value f determines the amount of flow that circulates back to vertex v_1 from v_m under this assumption. Since all flows are circulating and excess flow is neither created nor destroyed, the value of f is then the total flow that flows from v_1 to v_m. By maximizing f, Eq. (12.4) is exactly computing the maximum amount of flow that can go from vertex v_1 to v_m under the assumptions that flows are constrained by edge capacities ($\mathbf{x} \leq \mathbf{u}$), flows are non-negative ($\mathbf{x} \geq \mathbf{0}$), and flows are neither created nor destroyed in the graph.

12.2 The Dual of the Flow Maximization Problem

Theorem 12.10. *The dual linear programming problem for Eq.* (12.4) *is*

$$\begin{cases} \min & \displaystyle\sum_{k=1}^{n} u_k h_k \\ s.t. & w_m - w_1 = 1 \\ & w_i - w_j + h_k \geq 0 \quad \forall\, e_k = (v_i, v_j) \in E \\ & h_k \geq 0 \quad \forall\, (v_i, v_j) \in E \\ & w_i \textit{ unrestricted} \quad \forall i \in \{1, \ldots, m\}. \end{cases} \tag{12.5}$$

Proof. Consider the constraints of Eq. (12.4) and suppose that the imaginary edge from v_m to v_1 is edge e_0. We first add slack variables to constraints of the form $x_k \leq u_k$ to obtain

$$x_k + s_k = u_k \quad \forall k \in \{1, \dots, n\}.$$

The constraints (other than $\mathbf{x} \geq \mathbf{0}$ and f unrestricted) can be rewritten in matrix form as

$$\begin{bmatrix} -1 & a_{11} & a_{12} & \cdots & a_{1n} & 0 & 0 & \cdots & 0 \\ 0 & a_{12} & a_{22} & \cdots & a_{2n} & 0 & 0 & \cdots & 0 \\ \vdots & \vdots & \vdots & \vdots & \vdots & \vdots & \vdots & \vdots & \vdots \\ 1 & a_{m1} & a_{m2} & \cdots & a_{mn} & 0 & 0 & \cdots & 0 \\ 0 & 1 & 0 & \cdots & 0 & 1 & 0 & \cdots & 0 \\ 0 & 0 & 1 & \cdots & 0 & 0 & 1 & \cdots & 0 \\ \vdots & \vdots & \vdots & \vdots & \vdots & \vdots & \vdots & \vdots & \vdots \\ 0 & 0 & 0 & \cdots & 1 & 0 & 0 & \cdots & 1 \end{bmatrix} \begin{bmatrix} f \\ x_1 \\ x_2 \\ \vdots \\ x_n \\ s_1 \\ s_2 \\ \vdots \\ s_n \end{bmatrix} = \begin{bmatrix} 0 \\ 0 \\ \vdots \\ 0 \\ u_1 \\ u_2 \\ \vdots \\ u_n \end{bmatrix}. \tag{12.6}$$

or more simply as

$$\begin{bmatrix} \mathbf{e}_m - \mathbf{e}_1 & \mathbf{A} & \mathbf{0} \\ \mathbf{0} & \mathbf{I}_n & \mathbf{I}_n \end{bmatrix} \begin{bmatrix} f \\ \mathbf{x} \\ \mathbf{s} \end{bmatrix} = \begin{bmatrix} \mathbf{0} \\ \mathbf{u} \end{bmatrix}, \tag{12.7}$$

where all elements, written as $\mathbf{0}$, are zero matrices or vectors of appropriate dimension. This matrix has $2n+1$ columns and $m+n$ rows. To the first m rows, we associate the dual variables $w_1, \dots, w_m$. To the next n rows, we associate the dual variables $h_1, \dots, h_n$. Our dual variable vector is then

$$\mathbf{y} = [w_1, \dots, w_m, h_1, \dots, h_n].$$

Written in matrix form, this is

Constraints $\qquad\qquad$ Dual variables

$$\begin{bmatrix} \mathbf{e}_m - \mathbf{e}_1 & \mathbf{A} & \mathbf{0} \\ \mathbf{0} & \mathbf{I}_n & \mathbf{I}_n \end{bmatrix} \begin{bmatrix} f \\ \mathbf{x} \\ \mathbf{s} \end{bmatrix} = \begin{bmatrix} \mathbf{0} \\ \mathbf{u} \end{bmatrix} \qquad \begin{bmatrix} \mathbf{w} \\ \mathbf{h} \end{bmatrix}.$$

Since the constraints in Eq. (12.6) are all equality constraints, we know that these dual variables are unrestricted. From Eq. (12.4), we know that the objective function vector is

$$\mathbf{c} = [1, 0, \ldots, 0].$$

We can now compute our dual constraints. We use Eq. (11.26) and Fig. 11.4. The left-hand side of the dual constraints is computed as

$$\begin{bmatrix}\mathbf{w}^T & \mathbf{h}^T\end{bmatrix}\begin{bmatrix}\mathbf{e}_m - \mathbf{e}_1 & \mathbf{A} & \mathbf{0} \\ \mathbf{0} & \mathbf{I}_n & \mathbf{I}_n\end{bmatrix}. \tag{12.8}$$

We fill in the right-hand sides using the vector $\mathbf{c}$, and we use Fig. 11.4 with the primal variables $(f, \mathbf{x})$.

Multiplying the matrices in Eq. (12.8) we obtain the first dual constraint

$$-w_1 + w_m = 1. \tag{12.9}$$

Since this dual constraint corresponds to the variable f in the primal problem, we know it will be an equality constraint (f is unrestricted) and that its right-hand side will be 1, the coefficient of f in the primal problem. The next n constraints are derived similarly and correspond to the variables $x_1 \geq 0, \ldots, x_m \geq$ and so will be the inequality constraints

$$w_i - w_j + h_k \geq 0. \tag{12.10}$$

This follows since there will be a -1 in the matrix whenever edge e_k has as destination vertex v_j and a $+1$ in the matrix whenever edge e_k has source at vertex v_i. Clearly, there is a 1 in the kth row of the identity matrix below the $\mathbf{A}$ matrix in Eq. (12.6), thus yielding the $+h_k$ term. The final n constraints have the form

$$h_k \geq 0 \tag{12.11}$$

and are derived from the last n columns of the matrix in Eq. (12.6). These constraints correspond to the variables $s_1 \geq 0, \ldots, s_n \geq 0$. The objective function of the dual problem is computed as

$$\begin{bmatrix}\mathbf{w}^T & \mathbf{h}^T\end{bmatrix}\begin{bmatrix}\mathbf{u} \\ \mathbf{0}\end{bmatrix}.$$

This yields the objective function

$$\sum_{k=1}^{n} u_k h_k. \tag{12.12}$$

Equation (12.5) follows at once. This completes the proof. □

12.3 The Max-Flow/Min-Cut Theorem

Remark 12.11. Recall from Remark 6.10 that we define a vertex cut using two sets $V_1 \subset V$ and $V_2 = V \setminus V_1$ with $v_1 \in V_1$ and $v_m \in V_2$. This cut is referred to as (V_1, V_2) and consists of all edges connecting a vertex in V_1 to a vertex in V_2. It has a capacity of $C(V_1, V_2)$, given in Definition 6.11.

Remark 12.12. Unlike in Chapter 6, we use linear programming duality to obtain the following results.

Lemma 12.13. *Let $G = (V, E)$ be a directed graph, and suppose $V = \{v_1, \dots, v_m\}$ and $E = \{e_1, \dots, e_n\}$. The solution to the maximum flow problem is bounded above by the minimal cut capacity.*

Proof. Let (V_1, V_2) be the cut with minimal capacity. Consider the following solution to the dual problem:

$$w_i^* = \begin{cases} 0 & v_i \in V_1 \\ 1 & v_i \in V_2. \end{cases} \tag{12.13}$$

and

$$h_k^* = \begin{cases} 1 & e_k = (v_i, v_j) \quad \text{and} \quad v_i \in V_1 \quad \text{and} \quad v_j \in V_2, \\ 0 & \text{otherwise}. \end{cases} \tag{12.14}$$

It is clear that this represents a feasible solution to the dual problem. Thus, by the strong duality theorem (Theorem 11.32), the objective function value

$$\sum_{k} u_k h_k^*. \tag{12.15}$$

is an upper bound for the primal problem. But this is just the capacity of the cut with the smallest capacity. This completes the proof. □

Lemma 12.14. *In any optimal solution to Eq.* (12.4), *every directed path from v_1 to v_m must have at least one edge at capacity.*

Proof. Note first that Eq. (12.4) is bounded above by the capacity of the minimal cut, as shown in Lemma 12.13, and since the zero flow is a feasible solution, we know from Theorem 11.18 that there is at least one optimal solution to Eq. (12.4) because the problem can neither be unbounded nor infeasible.

Consider any optimal solution to Eq. (12.5). Then, it corresponds to some optimal solution to the primal problem, and these solutions satisfy the Karush–Kuhn–Tucker conditions. We show that in this primal solution, along each path from v_1 to v_m in G, at least one edge must have flow equal to its capacity. To see this, note that for any edge that does not carry its capacity (that is, $x_k < u_k$) we must have $h_k = 0$ (to ensure complementary slackness). Suppose this path has vertices $(u_1, u_2, \ldots, u_s)$ with $v_1 = u_1$ and $v_m = u_s$. If there is some path from v_1 to v_m that does not carry its capacity, then we have the following requirements:

$$
\begin{aligned}
w_s &> w_1, \\
w_1 &\geq w_2, \\
&\vdots \\
w_{s-1} &\geq w_s.
\end{aligned}
$$

However, this implies that $w_s > w_1 \geq w_2 \geq \cdots \geq w_s$, which is a contradiction. Therefore, every path from v_1 to v_m has at least one edge at capacity. □

Remark 12.15. The proofs of the next two results are identical to the ones in Chapter 6. See Theorem 6.15 and its corollary.

Theorem 12.16. *Let $G = (V, E)$ be a directed graph and suppose $V = \{v_1, \ldots, v_m\}$ and $E = \{e_1, \ldots, e_n\}$. There is at least one cut (V_1, V_2) so that the flow from v_1 to v_m is equal to the capacity of the cut (V_1, V_2).* □

Corollary 12.17 (Max-flow/Min-cut theorem). *Let $G = (V, E)$ be a directed graph, and suppose $V = \{v_1, \ldots, v_m\}$ and $E = \{e_1, \ldots, e_n\}$. Then, the maximum flow from v_1 to v_m is equal to the capacity of the minimum cut separating v_1 from v_m.* □

Remark 12.18. The derivation of the Ford–Fulkerson algorithm for finding a maximum flow and its proof of correctness do not require the linear programming formulation and thus can be found in Chapter 6. In the remainder of this chapter, we briefly discuss the min-cost flow problem and its computational complexity and the relationship between the primal and dual problems in König's theorem.

12.4 Min-Cost Flow and Other Problems

Definition 12.19 (Min-cost flow problem). Let $\mathbf{A}$ be the incidence matrix of a directed graph. Assume that flow x_k has cost $c_k \in \mathbb{R}$ and that the flow on edge k is constrained so that $l_k \leq x_k \leq u_k$, where $l_k \geq 0$ is a lower bound on the flow. If $b_i \in \mathbb{R}$ is the flow produced (or consumed) at vertex i, then the minimum cost flow problem is

$$\begin{aligned} \min\ & \mathbf{c}^T\mathbf{x} \\ s.t.\ & \mathbf{A}\mathbf{x} = \mathbf{b} \\ & \mathbf{l} \leq \mathbf{x} \leq \mathbf{u}, \end{aligned} \tag{12.16}$$

where $\mathbf{c}$, $\mathbf{b}$, $\mathbf{l}$, and $\mathbf{u}$ are vectors of the corresponding parameters. Then, Eq. (12.16) is a *minimum-cost flow problem.*

Remark 12.20. Solving problems of this type are outside the scope of this text, but they are discussed in Ref. [29], with the *network simplex algorithm* being one very efficient mechanism of solution. The following proposition follows from the fact that linear programs can be solved in polynomial time [121].

Proposition 12.21. *The minimum-cost flow problem can be solved in polynomial time.* □

Remark 12.22. Interestingly, these problems can be solved in polynomial time *even if we require the solution to be in integers.* This is *not* true of general linear programming problems and is what is known as being *strongly polynomial* [124]. We have already seen an example of this in the integer flow theorem (Corollary 6.31). More details on why this is the case can be found in Ref. [29]. As a result of this property, these minimum-cost flow problems find a variety of uses, which we illustrate in the following.

Example 12.23 (Assignment problems). Consider a group of m people who are to be assigned to n projects. Create a vertex for each person and project and an extra "dummy" project. Create a complete directed bipartite graph from people to projects. (We assume all people can work on all projects. If this is not the case, simply remove those edges.) To each person vertex, connect a source vertex, and from each project, connect a sink vertex. This is illustrated in Fig. 12.1. Flow in this graph will be hours assigned from a person to a project. Table 12.1 shows the parameters assigned to each edge. The cost per hour can be specific to an individual and project so that each person may cost a different amount on each project. The upper bounds on the edges from the sink to the people prevent individuals from being overscheduled. The lower bounds on the edges from the projects to the sink ensure each project is adequately staffed. The

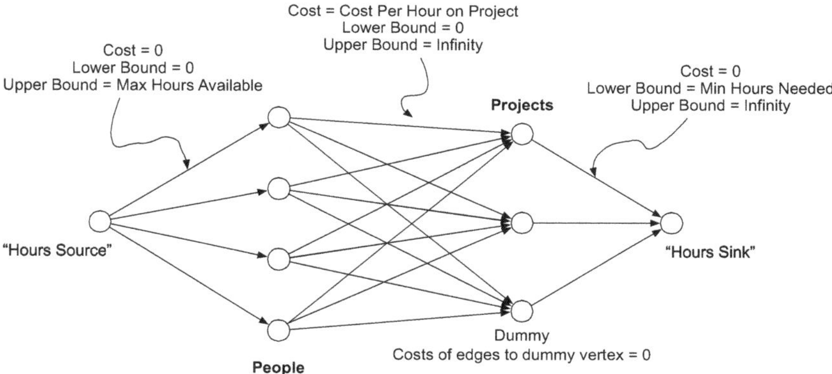

Fig. 12.1 A minimum cost flow problem that will solve a project assignment problem.

Table 12.1 A table showing the edge parameters for a simple assignment problem.

Edge Type	Lower Bound	Upper Bound	Cost
Source to person	0	Max. hours available	0
Person to project	0	∞	Cost per hour
Person to dummy	0	∞	0
Project to sink	Min. hours needed	∞	0
Dummy to sink	0	∞	0

Table 12.2 A table showing the vertex flow production.

Vertex Type	Flow Production (Consumption)
Source	Total person-hours available
Person or project	0
Sink	(Total person-hours available)

dummy collects excess hours that will not be used to schedule people to projects. The vertex flow production is given in Table 12.2. The source and sink generate (consume) flow, so they will have positive (negative) flow production values. All other vertices neither consume nor generate flow. Solving the minimum-cost flow problem will yield an assignment (in hours) of people to projects. If the problem is infeasible, it means that hour quotas cannot be met; i.e., projects demand too many hours. If the problem is feasible, then this is the cheapest possible assignment of people to projects.

Remark 12.24. Many assignment problems can be rephrased in this format or have (at their core) a flow problem that can be exploited to find solutions quickly. The interested reader can consult Ref. [29].

Example 12.25. Consider a simple assignment problem with three people and two projects. Both projects require 20 hours of work. We will call them Project 1 and Project 2:

(1) Alice costs \$20 per hour and is only qualified to work on Project 1. She can work a maximum of 20 hours.
(2) Bob costs \$15 per hour and can work on either project. He can work up to 30 hours.
(3) Charlie costs \$10 per hour and can work only on Project 1. He can work up to 10 hours.

The problem is illustrated in the network flow diagram in Fig. 12.2. Flow will leave Vertex 1 and go to Vertices 7 (the dummy vertex) and 8. The total number of hours available is 60 hours. This is the flow generated at Vertex 1. The projects absorb 40 hours ($b_1 = 40$). This is the flow absorbed at Vertex 8 ($b_8 = -40$). The dummy vertex (Vertex 7) absorbs the remaining 20 hours ($b_7 = -20$). Let $x_{ij} \geq 0$ be the amount of flow from Vertex i to Vertex j.

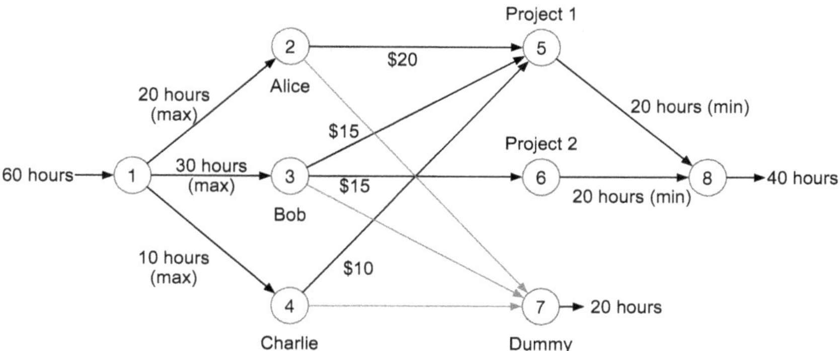

Fig. 12.2 An assignment problem with three workers and two projects. Constraints and costs are notionally illustrated on the edges.

The resulting linear programming problem is given by the expression

$$\begin{cases} \min \ 10x_{45} + 15x_{35} + 15x_{36} + 20x_{25} \\ s.t. \ x_{12} + x_{13} + x_{14} = 60 \\ x_{12} - x_{25} - x_{27} = 0 \\ x_{13} - x_{35} - x_{36} - x_{37} = 0 \\ x_{14} - x_{45} - x_{47} = 0 \\ x_{25} + x_{35} + x_{45} - x_{58} = 0 \\ x_{36} - x_{68} = 0 \\ x_{27} + x_{37} + x_{47} = 20 \\ x_{58} + x_{68} = 40 \\ x_{12} \leq 20 \\ x_{13} \leq 30 \\ x_{14} \leq 10 \\ x_{58} \geq 20 \\ x_{68} \geq 20 \\ x_{ij} \geq 0 \quad \forall i, j. \end{cases}$$

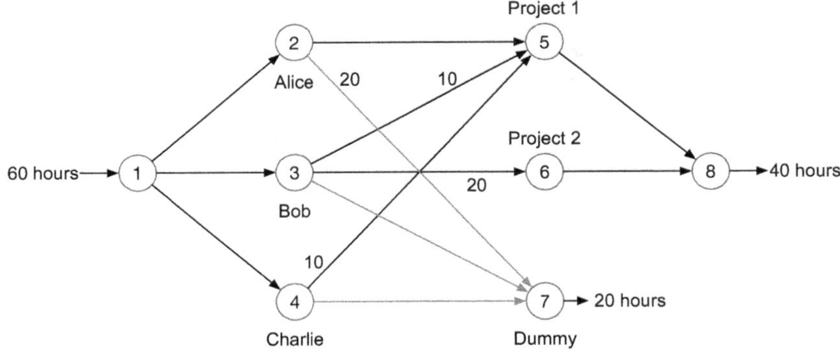

Fig. 12.3 The solution to assignment problem has Alice (the most expensive) not assigned to a project, while Bob and Charlie work on Projects 1 and 2.

Note that the incidence matrix of the graph in Fig. 12.2 is hidden in the flow conservation constraints of this problem.

Using a computer, we can conclude that the optimal solution has Bob assigned to Project 1 for 10 hours and Project 2 for 20 hours. Charlie is assigned to Project 1 for 10 hours. Alice (the most expensive worker) is not assigned hours. All her hours flow to the dummy vertex. This is illustrated in Fig. 12.3.

12.5 The Problem of Generalizing König's Theorem and Duality

Remark 12.26. In Chapter 6, we noted that König's theorem does not hold in general. To see this, consider K_3 (see Fig. 12.4). In this case, the general inequality that that the cardinality of the maximal matching is at most the cardinality of the minimal covering does hold (and this will always hold), but we do not have equality.

Remark 12.27. Let $G = (V, E)$ be a (bipartite) graph with $V = \{v_1, \ldots, v_m\}$ and $E = \{e_1, \ldots, e_n\}$. The minimal vertex covering problem for G can be written as the integer programming problem

$$\begin{cases} \min \ x_1 + \cdots + x_m \\ s.t. \ x_i + x_j \geq 1 \quad \forall \{v_i, v_j\} \in E \\ \qquad x_i \in \{0, 1\} \quad \forall i = 1, \ldots, m. \end{cases} \tag{12.17}$$

Fig. 12.4 In general, the cardinality of a maximal matching is not the same as the cardinality of a minimal vertex covering, though the inequality that the cardinality of the maximal matching is at most the cardinality of the minimal covering does hold.

Here, x_i acts as a boolean (true/false) variable that determines whether v_i is in the cover or not. If $\mathbf{A}$ is the incidence matrix for G, then this problem can be written in matrix notation as

$$\begin{cases} \min \ \mathbf{1}^T\mathbf{x} \\ s.t. \ \mathbf{A}^T\mathbf{x} \geq \mathbf{1} \\ \quad \mathbf{x} \in \{0,1\}^m, \end{cases} \tag{12.18}$$

where $\mathbf{1}$ is a vector consisting of only ones of appropriate length.

For simplicity, consider the *relaxation* of the integer program to a linear program:

$$\begin{cases} \min \ \mathbf{1}^T\mathbf{x} \\ s.t. \ \mathbf{A}^T\mathbf{x} \geq \mathbf{1} \\ \quad \mathbf{0} \leq \mathbf{x} \leq \mathbf{1}. \end{cases} \tag{12.19}$$

We can rewrite this as

$$\begin{cases} \min \ \mathbf{1}^T\mathbf{x} & \text{Dual Variables} \\ s.t. \ \begin{bmatrix} \mathbf{A}^T \\ -\mathbf{I}^T \end{bmatrix} \mathbf{x} \geq \begin{bmatrix} \mathbf{1} \\ -\mathbf{1} \end{bmatrix} & \begin{bmatrix} \mathbf{w} \\ \mathbf{u} \end{bmatrix} \\ \quad \mathbf{x} \geq \mathbf{0}. \end{cases}$$

The dual variables are shown to the right of the problem and $\mathbf{I}^T = \mathbf{I}$ because it's an identity matrix. The dual problem can be read from

Fig. 11.4:

$$\begin{cases} \max & \mathbf{1}^T\mathbf{w} - \mathbf{1}^T\mathbf{u} \\ s.t. & \mathbf{A}\mathbf{w} - \mathbf{I}\mathbf{u} \leq \mathbf{1} \\ & \mathbf{w}, \mathbf{u} \geq \mathbf{0}. \end{cases}$$

Remember that we are dealing with a dual problem generated from a *relaxation* of an integer programming problem of interest. To recover a sensible integer program from the dual, we note that setting $\mathbf{u} = \mathbf{0}$ will make the objective function larger. If we make that assumption, then we know that $\mathbf{w} \leq 1$ because of the structure of the constraints, and we can write

$$\begin{cases} \max & \mathbf{1}^T\mathbf{w} \\ s.t. & \mathbf{A}\mathbf{w} \leq \mathbf{1} \\ & \mathbf{0} \leq \mathbf{w} \leq \mathbf{1}. \end{cases}$$

Transforming this back to an integer program yields

$$\begin{cases} \max & \mathbf{1}^T\mathbf{w} \\ s.t. & \mathbf{A}\mathbf{w} \leq \mathbf{1} \\ & \mathbf{w} \in \{0, 1\}^n, \end{cases}$$

which is the maximum matching problem. Because we know that relaxations will always have optimal objective functions that are at least as good as their corresponding integer programming problems, we have proved that

$$\mathrm{IP}_{\text{match}} \leq \mathrm{LP}_{\text{match}} \leq \mathrm{LP}_{\text{cover}} \leq \mathrm{IP}_{\text{cover}}. \tag{12.20}$$

The middle inequality follows from weak duality and our assumption on the variables $\mathbf{u}$. However, we have seen for the special case of bipartite graphs that these inequalities must be equalities. This is the König's theorem.

12.6 Chapter Notes

The interplay between graph structures and linear and integer programming has led to interesting theoretical and practical results.

The minimum-cost flow problem can be solved in a number of different ways, including the network simplex method, the out-of-kilter algorithm [125], and the push-relabel algorithm [126]. There is no single original network simplex paper, though the idea can be traced as far back as Koopmans [127], with major work on the polynomial method by Orlin, Plotkin, and Tardos [128], Orlin [129], and Trajan [130]. The out-of-kilter algorithm was developed by Fulkerson [125] (the same Fulkerson of the Ford–Fulkerson algorithm) and is unique in that it obtains feasible solutions to both the primal and dual linear programming problems underlying the minimum-cost flow problem, but these solutions do not satisfy complementary slackness. The algorithm proceeds to correct this problem, ultimately arriving at a point satisfying the KKT conditions. See Ref. [29] for a complete description. The push-relabel algorithm is the current fastest solution method for maximum flow problems with a running time of $O(|V|^2|E|)$ compared to Theorem 6.29, which gives a running time of $O(|V||E|^2)$ for the Edmonds–Karp algorithm. Interestingly, Dijkstra's algorithm can even be derived from the dual problem of an appropriately phrased linear programming problem. An alternative proof of that algorithm's correctness uses the KKT conditions, just as we did for the maximum flow problem in this chapter. It is worth noting that many of these citations occur from 1988 onward, showing that this is a recent area of active research.

The general matching problem (and its generalization, the weighted matching problem) on arbitrary graphs is also a much studied area of combinatorial optimization. The blossom algorithm developed by Edmonds [131] (of the Edmonds–Karp algorithm) is known to solve this problem in polynomial time $O(|V||E|^2)$. This is a second example of a combinatorial optimization problem that can be phrased as an integer programming problem that can be solved in polynomial time. (Integer programming problems, in general, cannot be solved in polynomial time unless P $=$ NP, which is still open.) The fact that this is the case is important because it arises from a different mechanism than the polynomial time complexity of the network flow problem. As a result, the so-called matching polytope [132], describing the feasible region of the matching problem, has been much studied [133] because of this polynomial time property. Works dedicated to integer and linear programming [29, 59, 115, 134] will have additional information on the relationship between graphs and networks and linear programming.

There are many other applications of linear programming (and integer programming) to the study of graphs. Please consult Refs. [29, 59, 115, 134].

12.7 Exercises

Exercise 12.1
Show that the maximum flow problem can be rephrased as a minimum-cost flow problem. Consequently, prove that it is a special case of a minimum-cost flow problem.

Exercise 12.2
For the graph K_3, find the integer programming problems and its relaxation for the minimum covering and maximum matching problems. Find optimal solutions to each problem, and illustrate Eq. (12.20).

Exercise 12.3
Remove one edge from K_3 to form a bipartite graph and repeat Question 12.2. Illustrate the König's theorem in this case with the integer/linear programs.

Appendix A

Fields, Vector Spaces, and Matrices

Remark A.1. This appendix introduces the essentials of linear algebra needed for the results on algebraic graph theory presented in the main part of the book. Proofs are omitted for brevity. Reading Chapter 8 will be helpful for understanding the definition of *group*, which is used throughout this appendix. The Perron–Frobenius theorem, a useful result in linear algebra, is not presented in this chapter. Instead, it is presented where it is needed in the main body of the text.

A.1 Matrices and Row and Column Vectors

Definition A.2 (Field). A *field* (or number field) is a tuple $(S, +, \cdot, 0, 1)$ where:

(1) $(S, +)$ is a commutative group with unit 0;
(2) $(S \setminus \{0\}, \cdot)$ is a commutative group with unit 1;
(3) the operation $\cdot$ *distributes* over the operation $+$ so that if a_1, a_2, and a_3 are elements of F, then $a_1 \cdot (a_2 + a_3) = a_1 \cdot a_2 + a_1 \cdot a_3$.

Example A.3. The archetypal example of a field is the field of real numbers $\mathbb{R}$, with addition and multiplication playing the expected roles. Another common field is the field of complex numbers $\mathbb{C}$ (numbers of the form $a + bi$ with $i = \sqrt{-1}$ the imaginary unit), with their addition and multiplication rules defined as expected.

Definition A.4 (Matrix). An $m \times n$ matrix is a rectangular array of values (*scalars*) drawn from a field. If $\mathbb{F}$ is the field, we write $\mathbb{F}^{m\times n}$ to denote the set of $m \times n$ matrices with entries drawn from $\mathbb{F}$.

Example A.5. Here is an example of a 2×3 matrix drawn from $\mathbb{R}^{2\times 3}$:

$$\mathbf{A} = \begin{bmatrix} 3 & 1 & \frac{7}{2} \\ 2 & \sqrt{2} & 5 \end{bmatrix}.$$

Remark A.6. We denote the element at position (i, j) of matrix $\mathbf{A}$ as $\mathbf{A}_{ij}$. Thus, in the example above, $\mathbf{A}_{2,1} = 2$.

Definition A.7 (Matrix addition). If $\mathbf{A}$ and $\mathbf{B}$ are both in $\mathbb{F}^{m\times n}$, then $\mathbf{C} = \mathbf{A} + \mathbf{B}$ is the matrix sum of $\mathbf{A}$ and $\mathbf{B}$ in $\mathbb{F}^{m\times n}$ and

$$\mathbf{C}_{ij} = \mathbf{A}_{ij} + \mathbf{B}_{ij} \quad \text{for } i = 1, \ldots, m \text{ and } j = 1, \ldots, n. \tag{A.1}$$

Here, $+$ is the field operation for addition.

Example A.8.

$$\begin{bmatrix} 1 & 2 \\ 3 & 4 \end{bmatrix} + \begin{bmatrix} 5 & 6 \\ 7 & 8 \end{bmatrix} = \begin{bmatrix} 1+5 & 2+6 \\ 3+7 & 4+8 \end{bmatrix} = \begin{bmatrix} 6 & 8 \\ 10 & 12 \end{bmatrix}. \tag{A.2}$$

Definition A.9 (Scalar-matrix multiplication). If $\mathbf{A}$ is a matrix from $\mathbb{F}^{m\times n}$ and $c \in \mathbb{F}$, then $\mathbf{B} = c\mathbf{A} = \mathbf{A}c$ is the scalar-matrix product of c and $\mathbf{A}$ in $\mathbb{F}^{m\times n}$ and

$$\mathbf{B}_{ij} = c\mathbf{A}_{ij} \quad \text{for } i = 1, \ldots, m \quad \text{and} \quad j = 1, \ldots, n. \tag{A.3}$$

Example A.10. Let

$$\mathbf{B} = \begin{bmatrix} 3 & 7 \\ 6 & 3 \end{bmatrix}.$$

When we multiply by the scalar $3 \in \mathbb{R}$, we obtain

$$3 \cdot \begin{bmatrix} 3 & 7 \\ 6 & 3 \end{bmatrix} = \begin{bmatrix} 9 & 21 \\ 18 & 9 \end{bmatrix}.$$

Definition A.11 (Row/Column vector). A $1 \times n$ matrix is called a *row vector*, and a $m \times 1$ matrix is called a *column vector*. Every vector will be thought of as a **column vector** unless otherwise noted. A column vector $\mathbf{x}$ in $\mathbb{R}^{n\times 1}$ (or $\mathbb{R}^n$) is: $\mathbf{x} = \langle x_1, \ldots, x_n \rangle$.

Remark A.12. It should be clear that any row of matrix $\mathbf{A}$ could be considered a row vector and any column of $\mathbf{A}$ could be considered a column vector. The ith row of $\mathbf{A}$ is denoted as $\mathbf{A}_{i\cdot}$, while the jth column is denoted as $\mathbf{A}_{\cdot j}$. Also, any row/column vector is nothing more sophisticated than tuples of numbers (a point in space). You are free to think of these things however you like.

Definition A.13 (n-dimensional vector space $\mathbb{F}^n$). Let $\mathbb{F}$ be a field (e.g., the real numbers). The set of all vectors $\mathbf{x} = \langle x_1, \ldots, x_n \rangle$ with $x_i \in \mathbb{F}$ is the *vector space* $\mathbb{F}^n$. The *dimension* of $\mathbb{F}^n$ is n.

Proposition A.14. *The vector space $\mathbb{F}^n$ is a group under vector addition, with the zero vector $\mathbf{0} = \langle 0, 0, \ldots, 0 \rangle$ being the additive identity. Furthermore, $\mathbb{F}^n$ is closed under scalar multiplication.* □

Remark A.15. The vector space $\mathbb{F}^n$ is usually the first one students learn. In particular, we speak of $\mathbb{R}^n$ in most classes on matrices. In linear algebra, we learn that more abstract things can be made into vector spaces, all of which are isomorphic to some vector space $\mathbb{F}^n$. For the purpose of this book, we confine ourselves to these archetypal vector spaces.

A.2 Matrix Multiplication

Definition A.16 (Dot product). If $\mathbf{x}, \mathbf{y} \in \mathbb{F}^n$ are two n-dimensional vectors, then their *dot product* is

$$\mathbf{x} \cdot \mathbf{y} = \sum_{i=1}^{n} x_i y_i, \tag{A.4}$$

where x_i is the ith component of the vector $\mathbf{x}$.

Remark A.17. The dot product is an example of a more general concept called an *inner product*, which maps two vectors to a scalar. Not all inner products behave according to Eq. (A.4), and the definition of the inner product will be vector-space- and context-dependent.

Definition A.18 (Matrix multiplication). If $\mathbf{A} \in \mathbb{R}^{m\times n}$ and $B \in \mathbb{R}^{n\times p}$, then $\mathbf{C} = \mathbf{AB}$ is the *matrix product* of $\mathbf{A}$ and $\mathbf{B}$ and

$$\mathbf{C}_{ij} = \mathbf{A}_{i\cdot} \cdot \mathbf{B}_{\cdot j}. \tag{A.5}$$

Note that $\mathbf{A}_{i\cdot} \in \mathbb{R}^{1\times n}$ (an n-dimensional vector) and $\mathbf{B}_{\cdot j} \in \mathbb{R}^{n\times 1}$ (another n-dimensional vector), thus making the dot product meaningful. Note also that $\mathbf{C} \in \mathbb{R}^{m\times p}$.

Example A.19.

$$\begin{bmatrix} 1 & 2 \\ 3 & 4 \end{bmatrix} \begin{bmatrix} 5 & 6 \\ 7 & 8 \end{bmatrix} = \begin{bmatrix} 1(5)+2(7) & 1(6)+2(8) \\ 3(5)+4(7) & 3(6)+4(8) \end{bmatrix} = \begin{bmatrix} 19 & 22 \\ 43 & 50 \end{bmatrix}. \tag{A.6}$$

Remark A.20. Note that we cannot multiply any pair of arbitrary matrices. If we have the product $\mathbf{AB}$ for two matrices $\mathbf{A}$ and $\mathbf{B}$, then the number of columns in $\mathbf{A}$ must be equal to the number of rows in $\mathbf{B}$.

Definition A.21 (Matrix transpose). If $\mathbf{A} \in \mathbb{F}^{m\times n}$ is a $m \times n$ matrix, then the *transpose* of $\mathbf{A}$, denoted by $\mathbf{A}^T$, is an $m \times n$ matrix defined as

$$\mathbf{A}^T_{ij} = \mathbf{A}_{ji}. \tag{A.7}$$

Example A.22.

$$\begin{bmatrix} 1 & 2 \\ 3 & 4 \end{bmatrix}^T = \begin{bmatrix} 1 & 3 \\ 2 & 4 \end{bmatrix}. \tag{A.8}$$

Essentially, we are just reading down the columns of $\mathbf{A}$ to obtain its transpose.

Remark A.23. The matrix transpose is a particularly useful operation and makes it easy to transform column vectors into row vectors, which enables multiplication. For example, suppose $\mathbf{x}$ is an $n \times 1$ column vector (i.e., $\mathbf{x}$ is a vector in $\mathbb{F}^n$), and suppose $\mathbf{y}$ is an $n \times 1$ column vector. Then,

$$\mathbf{x} \cdot \mathbf{y} = \mathbf{x}^T\mathbf{y}. \tag{A.9}$$

A.3 Special Matrices

Definition A.24 (Square matrix). Any matrix $\mathbf{A} \in \mathbb{F}^{n\times n}$ for a field $\mathbb{F}$ is called a *square matrix.*

Remark A.25. There are many special (and useful) square matrices, as we discuss in the following.

Definition A.26 (Identity matrix). The $n \times n$ *identify matrix* is

$$\mathbf{I}_n = \begin{bmatrix} 1 & 0 & \dots & 0 \\ 0 & 1 & \dots & 0 \\ \vdots & & \ddots & \vdots \\ 0 & 0 & \dots & 1 \end{bmatrix}. \tag{A.10}$$

Here, 1 is the multiplicative unit in the field $\mathbb{F}$ from which the matrix entries are drawn.

Definition A.27 (Zero matrix). The $n \times n$ *zero* matrix is a $n \times n$ consisting entirely of 0's (the zero in the field).

Definition A.28 (Symmetric matrix). Let $\mathbf{M} \in \mathbb{F}^{n\times n}$ be a matrix. The matrix $\mathbf{M}$ is symmetric if $\mathbf{M} = \mathbf{M}^T$.

Example A.29. Suppose that

$$\mathbf{A} = \begin{bmatrix} 1 & 2 & 3 \\ 2 & 4 & 1 \\ 3 & 1 & 7 \end{bmatrix}.$$

This matrix is symmetric.

Definition A.30 (Diagonal matrix). A *diagonal matrix* is a (square) matrix with the property that $\mathbf{D}_{ij} = 0$ for $i \neq j$m and $\mathbf{D}_{ii}$ may take any value in the field on which $\mathbf{D}$ is defined.

Example A.31. Consider the matrix

$$\mathbf{D} = \begin{bmatrix} 2 & 0 & 0 \\ 0 & 4 & 0 \\ 0 & 0 & 6 \end{bmatrix}.$$

This is a diagonal matrix.

Remark A.32. A diagonal matrix has (usually) nonzero entries only on its main diagonal. Of course some of its diagonal entries may be zero.

A.4 Matrix Inverse

Definition A.33 (Invertible matrix). Let $\mathbf{A} \in \mathbb{F}^{n\times n}$ be a square matrix. If there is a matrix $\mathbf{A}^{-1}$ such that

$$\mathbf{A}\mathbf{A}^{-1} = \mathbf{A}^{-1}\mathbf{A} = \mathbf{I}_n, \tag{A.11}$$

then matrix $\mathbf{A}$ is said to be *invertible* (or *nonsingular*) and $\mathbf{A}^{-1}$ is called its *inverse*. If $\mathbf{A}$ is not invertible, it is called a *singular* matrix.

Definition A.34 (Matrix power). If $\mathbf{A} \in \mathbb{F}^{n\times n}$, then $\mathbf{A}^k = \mathbf{A}^{k-1}\mathbf{A} = \mathbf{A}\mathbf{A}^{k-1}$ for a positive integer k. We define $\mathbf{A}^0 = \mathbf{I}_n$ for any nonsingular square matrix $\mathbf{A}$.

Theorem A.35. *If $\mathbf{A} \in \mathbb{F}^{n\times n}$ is invertible, then $\mathbf{A}^{-1}$ is unique.* □

Remark A.36. Definition A.33 and Theorem A.35 show that the inverse of a square matrix is unique if it exists and there is no difference between a *left* inverse and a *right* inverse.

Remark A.37. The set of $n \times n$ invertible matrices over $\mathbb{R}$ is denoted by $\mathrm{GL}(n, \mathbb{R})$. It forms a group called the *general linear group* under matrix multiplication, with $\mathbf{I}_n$ as the unit. The general linear group over a field $\mathbb{F}$ is defined analogously.

Proposition A.38. *If both $\mathbf{A}$ and $\mathbf{B}$ are invertible in $\mathbb{F}^{n\times n}$, then $\mathbf{AB}$ is invertible and $(\mathbf{AB})^{-1} = \mathbf{B}^{-1}\mathbf{A}^{-1}$.* □

A.5 Linear Combinations, Span, and Linear Independence

Definition A.39. Let $\mathbf{v}_1, \ldots, \mathbf{v}_m$ be (column) vectors in $\mathbb{F}^n$, and let $\alpha_1, \ldots, \alpha_m \in \mathbb{F}$ be scalars. Then,

$$\alpha_1\mathbf{v}_1 + \cdots + \alpha_m\mathbf{v}_m \tag{A.12}$$

is a *linear combination* of the vectors $\mathbf{v}_1, \ldots, \mathbf{v}_m$.

Definition A.40 (Span). Suppose $W = \{\mathbf{v}_1, \ldots, \mathbf{v}_m\}$ is a set of vectors in $\mathbb{F}^n$, then the span of W is the set

$$\text{span}(W) = \{\mathbf{y} \in \mathbb{F}^n : \mathbf{y} \text{ is a linear combination of vectors in } W\}. \tag{A.13}$$

Definition A.41 (Linear independence). Let $\mathbf{v}_1, \ldots, \mathbf{v}_m$ be vectors in $\mathbb{F}^n$. The vectors $\mathbf{v}_1, \ldots, \mathbf{v}_m$ are *linearly dependent* if there exists $\alpha_1, \ldots, \alpha_m \in \mathbb{F}$, not all zero, such that

$$\alpha_1\mathbf{v}_1 + \cdots + \alpha_m\mathbf{v}_m = \mathbf{0}. \tag{A.14}$$

If the set of vectors $\mathbf{v}_1, \ldots, \mathbf{v}_m$ is not linearly dependent, then they are *linearly independent* and Eq. (A.14) holds just in case $\alpha_i = 0$ for all $i = 1, \ldots, n$. Here, $\mathbf{0}$ is the zero vector in $\mathbb{F}^n$ and 0 is the zero element in the field.

Example A.42. In $\mathbb{R}^3$, consider the vectors

$$\mathbf{v}_1 = \begin{bmatrix} 1 \\ 1 \\ 0 \end{bmatrix}, \quad \mathbf{v}_2 = \begin{bmatrix} 1 \\ 0 \\ 1 \end{bmatrix}, \quad \text{and} \quad \mathbf{v}_3 = \begin{bmatrix} 0 \\ 1 \\ 1 \end{bmatrix}.$$

We can show that these vectors are linearly independent: Suppose there are values $\alpha_1, \alpha_2, \alpha_3 \in \mathbb{R}$ such that

$$\alpha_1\mathbf{v}_1 + \alpha_2\mathbf{v}_2 + \alpha_3\mathbf{v}_3 = 0.$$

Then,

$$\begin{bmatrix} \alpha_1 \\ \alpha_1 \\ 0 \end{bmatrix} + \begin{bmatrix} \alpha_2 \\ 0 \\ \alpha_2 \end{bmatrix} \begin{bmatrix} 0 \\ \alpha_3 \\ \alpha_3 \end{bmatrix} = \begin{bmatrix} \alpha_1 + \alpha_2 \\ \alpha_1 + \alpha_3 \\ \alpha_2 + \alpha_3 \end{bmatrix} = \begin{bmatrix} 0 \\ 0 \\ 0 \end{bmatrix}.$$

Thus, we have the system of linear equations

$$\begin{aligned} \alpha_1 + \alpha_2 &= 0 \\ \alpha_1 + \alpha_3 &= 0 \\ \alpha_2 + \alpha_3 &= 0. \end{aligned}$$

From the third equation, we see that $\alpha_3 = -\alpha_2$. Substituting this into the second equation, we obtain two equations:

$$\alpha_1 + \alpha_2 = 0$$
$$\alpha_1 - \alpha_2 = 0.$$

This implies that $\alpha_1 = \alpha_2$ and $2\alpha_1 = 0$ or $\alpha_1 = \alpha_2 = 0$. Therefore, $\alpha_3 = 0$, and thus, these vectors are linearly independent.

Remark A.43. It is worth noting that any set of vectors becomes linearly dependent if the zero vector $\mathbf{0}$ is added to it.

Example A.44. Consider the vectors

$$\mathbf{v}_1 = \begin{bmatrix} 1 \\ 2 \\ 3 \end{bmatrix} \quad \text{and} \quad \mathbf{v}_2 = \begin{bmatrix} 4 \\ 5 \\ 6 \end{bmatrix}.$$

Determining linear independence requires us to find solutions to the equation

$$\alpha_1 \begin{bmatrix} 1 \\ 2 \\ 3 \end{bmatrix} + \alpha_2 \begin{bmatrix} 4 \\ 5 \\ 6 \end{bmatrix} = \begin{bmatrix} 0 \\ 0 \\ 0 \end{bmatrix}$$

or the system of equations

$$\alpha_1 + 4\alpha_2 = 0,$$
$$2\alpha_1 + 5\alpha_2 = 0,$$
$$3\alpha_1 + 6\alpha_2 = 0.$$

Thus, $\alpha_1 = -4\alpha_2$. Substituting this into the second and third equations yields

$$-3\alpha_2 = 0,$$
$$-6\alpha_2 = 0.$$

Thus, $\alpha_2 = 0$ and, consequently, $\alpha_1 = 0$. Thus, the vectors are linearly independent.

Example A.45. Consider the vectors

$$\mathbf{v}_1 = \begin{bmatrix} 1 \\ 2 \end{bmatrix}, \quad \mathbf{v}_2 = \begin{bmatrix} 3 \\ 4 \end{bmatrix}, \quad \text{and} \quad \mathbf{v}_3 = \begin{bmatrix} 5 \\ 6 \end{bmatrix}.$$

As before, we can derive the system of equations

$$\begin{aligned} \alpha_1 + 3\alpha_2 + 5\alpha_3 &= 0, \\ 2\alpha_1 + 4\alpha_2 + 6\alpha_3 &= 0. \end{aligned}$$

We have more unknowns than equations, so we suspect there may be many solutions to this system of equations. From the first equation, we see that $\alpha_1 = -3\alpha_2 - 5\alpha_3$. Consequently, we can substitute this into the second equation to obtain

$$-6\alpha_2 - 10\alpha_3 + 4\alpha_2 + 6\alpha_3 = -2\alpha_2 - 4\alpha_3 = 0.$$

Thus, $\alpha_2 = -2\alpha_3$ and $\alpha_1 = 6\alpha_3 - 5\alpha_3 = \alpha_3$, which we obtain by substituting the expression for α_2 into the expression for α_1. We conclude that α_3 can be anything we like; it is a free variable. Set $\alpha_3 = 1$. Then, $\alpha_2 = -2$ and $\alpha_1 = 1$. We can now confirm that this set of values creates a linear combination of $\mathbf{v}_1$, $\mathbf{v}_2$, and $\mathbf{v}_3$ equal to $\mathbf{0}$. We compute

$$1 \cdot \begin{bmatrix} 1 \\ 2 \end{bmatrix} - 2 \cdot \begin{bmatrix} 3 \\ 4 \end{bmatrix} + \begin{bmatrix} 5 \\ 6 \end{bmatrix} = \begin{bmatrix} 1 - 6 + 5 \\ 2 - 8 + 6 \end{bmatrix} = \begin{bmatrix} 0 \\ 0 \end{bmatrix}.$$

Thus, the vectors are *not* linearly independent and they must be linearly dependent.

A.6 Basis

Definition A.46 (Basis). Let $\mathcal{B} = \{\mathbf{v}_1, \ldots, \mathbf{v}_m\}$ be a set of vectors in $\mathbb{F}^n$. The set $\mathcal{B}$ is called a *basis* of $\mathbb{F}^n$ if $\mathcal{B}$ is a linearly independent set of vectors and every vector in $\mathbb{F}^n$ is in the span of $\mathcal{B}$. That is, for any vector $\mathbf{w} \in \mathbb{F}^n$, we can find scalar values $\alpha_1, \ldots, \alpha_m \in \mathbb{F}$ such that

$$\mathbf{w} = \sum_{i=1}^{m} \alpha_i \mathbf{v}_i. \tag{A.15}$$

Example A.47. We can show that the vectors

$$\mathbf{v}_1 = \begin{bmatrix} 1 \\ 1 \\ 0 \end{bmatrix}, \quad \mathbf{v}_2 = \begin{bmatrix} 1 \\ 0 \\ 1 \end{bmatrix}, \quad \text{and} \quad \mathbf{v}_3 = \begin{bmatrix} 0 \\ 1 \\ 1 \end{bmatrix}$$

form a basis of $\mathbb{R}^3$. We already know that the vectors are linearly independent. To show that every vector in $\mathbb{R}^3$ is in its span, chose an arbitrary vector in $\mathbb{R}^3$: $\langle a, b, c \rangle$. Then, we hope to find coefficients α_1, α_2, and α_3 so that

$$\alpha_1 \mathbf{v}_1 + \alpha_2 \mathbf{v}_2 + \alpha_3 \mathbf{v}_3 = \begin{bmatrix} a \\ b \\ c \end{bmatrix}.$$

Expanding this, we must find α_1, α_2, and α_3 so that

$$\begin{bmatrix} \alpha_1 \\ \alpha_1 \\ 0 \end{bmatrix} + \begin{bmatrix} \alpha_2 \\ 0 \\ \alpha_2 \end{bmatrix} + \begin{bmatrix} 0 \\ \alpha_3 \\ \alpha_3 \end{bmatrix} = \begin{bmatrix} a \\ b \\ c \end{bmatrix}.$$

A little effort (in terms of algebra) will show that

$$\begin{cases} \alpha_1 = \dfrac{1}{2}(a + b - c), \\ \alpha_2 = \dfrac{1}{2}(a - b + c), \\ \alpha_3 = \dfrac{1}{2}(-a + b + c). \end{cases} \tag{A.16}$$

Thus, the set $\{\mathbf{v}_1, \mathbf{v}_2, \mathbf{v}_3\}$ is a basis for $\mathbb{R}^3$.

Remark A.48. Note that there are three vectors in this basis for $\mathbb{R}^3$. In general, a basis for $\mathbb{F}^n$ will have exactly n vectors.

Definition A.49 (Standard basis). The standard basis of $\mathbb{F}^n$ consists of n columns (or rows) of the identity matrix $\mathbf{I}_n$. In most (but not all) texts, they are written as

$$\mathbf{e}_1 = \langle 1, 0, \ldots, 0 \rangle \quad \mathbf{e}_2 = \langle 0, 1, 0, \ldots, 0 \rangle \quad \cdots \quad \mathbf{e}_n = \langle 0, 0, \ldots, 1 \rangle.$$

A.7 Orthogonality in $\mathbb{R}^n$

Remark A.50. When two vectors are perpendicular to each other in some vector space, they are said to be *orthogonal*. We only need orthogonality in $\mathbb{R}^n$, so we formally define the idea for vectors with real entries. The definitions are similar but slightly different for vectors in, for example, $\mathbb{C}^n$. In particular, the dot product used in the following definition is replaced with a suitable inner product.

Definition A.51 (Orthogonality). Two vectors $\mathbf{x}$ and $\mathbf{y}$ in $\mathbb{R}^n$ are orthogonal if $\mathbf{x} \cdot \mathbf{y} = 0$.

Example A.52. It is easy to see that the vectors $\langle 1, 0\rangle$ and $\langle 0, 1\rangle$ are orthogonal.

Definition A.53 (Norm). The norm of a vector $\mathbf{x} \in \mathbb{R}^n$ is given by

$$\|\mathbf{x}\| = \sqrt{\mathbf{x} \cdot \mathbf{x}}.$$

Definition A.54 (Unit vector). A vector $\mathbf{x} \in \mathbb{R}^n$ is a unit vector if $\|\mathbf{x}\| = 1$.

Remark A.55. If we can define a vector norm in an arbitrary vector space $\mathbb{F}^n$, then any vector with norm 1 is called a unit vector. Even though we only need these ideas for vectors with real entries, it is nice to know that all of these definitions can be suitably generalized.

Definition A.56 (Orthogonal/Orthonormal basis). A basis $\mathcal{B}$ is *orthogonal* if all the vectors in it are mutually orthogonal to each other. It is orthonormal if all the vectors are mutually orthogonal and they are all unit vectors.

Remark A.57. It is easy to see that the standard basis is always orthonormal in $\mathbb{R}^n$. We will see another example of an orthonormal basis when we introduce eigenvalues in a later section.

A.8 Row Space and Null Space

Definition A.58 (Row space). Let $\mathbf{A} \in \mathbb{F}^{m \times n}$. The *row space* of $\mathbf{A}$ is the vector space $\mathbb{F}^k$ spanned by the rows of $\mathbf{A}$.

Example A.59. We have already shown in Example A.47 that the row space of the matrix

$$\mathbf{A} = \begin{bmatrix} 1 & 1 & 0 \\ 1 & 0 & 1 \\ 0 & 1 & 1 \end{bmatrix}$$

is the space $\mathbb{R}^3$.

Example A.60. Consider the matrix

$$\mathbf{A} = \begin{bmatrix} 1 & 2 \\ 3 & 4 \\ 5 & 6 \end{bmatrix}.$$

The rows come from $\mathbb{R}^2$. Consequently, we know that the row space is at most $\mathbb{R}^2$ (though it could be $\mathbb{R}^1$). By Remark A.48, we know that the row vectors must be linearly dependent (they are in $\mathbb{R}^2$ and there are three of them). Therefore, we show that two of the vectors are linearly independent, and consequently, they must span $\mathbb{R}^2$. To see this, we solve

$$\alpha_1 \begin{bmatrix} 1 & 2 \end{bmatrix} + \alpha_2 \begin{bmatrix} 3 & 4 \end{bmatrix} = \begin{bmatrix} 0 & 0 \end{bmatrix}.$$

Then, we have the equation system

$$\alpha_1 + 3\alpha_2 = 0,$$
$$2\alpha_1 + 4\alpha_2 = 0.$$

Multiplying the first equation by -2 and adding, we obtain

$$-2\alpha_2 = 0.$$

Therefore, $\alpha_2 = 0$, and it follows that $\alpha_1 = 0$. These two vectors are linearly independent. A similar argument shows that they span $\mathbb{R}^2$. Therefore, the row space of $\mathbf{A}$ is $\mathbb{R}^2$.

Definition A.61 (Rank). The rank of matrix $\mathbf{A} \in \mathbb{R}^{m\times n}$ is the dimension of the row space or, equivalently, the number of linearly independent rows of $\mathbf{A}$.

Definition A.62 (Null space). The *null space* of a matrix $\mathbf{A} \in \mathbb{F}^{m\times n}$ is the set of vectors $\mathcal{N}(\mathbf{A})$ such that if $\mathbf{x} \in \mathcal{N}(\mathbf{A})$, then $\mathbf{Ax} = \mathbf{0}$.

Proposition A.63. *If the columns of* $\mathbf{A}$ *are linearly independent, then the null space of* $\mathbf{A}$ *consists of only the zero vector (and has a dimension of zero).* □

Example A.64. Consider the matrix

$$\mathbf{A} = \begin{bmatrix} 1 & 2 & 3 \\ 2 & 4 & 6 \end{bmatrix}.$$

It should be clear that the rank of this matrix is 1 because row 2 is a multiple of row 1. We can construct the null space by solving

$$\begin{bmatrix} 1 & 2 & 3 \\ 2 & 4 & 6 \end{bmatrix} \begin{bmatrix} x_1 \\ x_2 \\ x_3 \end{bmatrix} = \begin{bmatrix} 0 \\ 0 \end{bmatrix}.$$

This leads to two equations:

$$x_1 + 2x_2 + 3x_3 = 0,$$
$$2x_1 + 4x_2 + 3x_3 = 0,$$

which are really just one equation. Then, setting

$$x_1 + 2x_2 + 3x_3 = 0$$

implies that

$$x_1 = -2x_2 + 3x_3.$$

We can set $x_2 = \alpha_1$ and $x_3 = \alpha_2$ as *free variables.* (Here, α_1 and α_2 may take on any values.) Then, the solutions to this system of equations have the form

$$\begin{bmatrix} x_1 \\ x_2 \\ x_3 \end{bmatrix} = \begin{bmatrix} -2\alpha_1 - 3\alpha_3 \\ \alpha_1 \\ \alpha_2 \end{bmatrix} = \alpha_1 \begin{bmatrix} -2 \\ 1 \\ 0 \end{bmatrix} + \alpha_2 \begin{bmatrix} -3 \\ 0 \\ 1 \end{bmatrix}.$$

This means that the space $\mathcal{N}(\mathbf{A})$ is spanned by the vectors $\langle -2, 1, 0 \rangle$ and $\langle -3, 0, 1 \rangle$. This is a *basis* for the null space. Consequently, since it has two basis vectors, it has a dimension of 2 and it can be put into one-to-one correspondence with $\mathbb{F}^2$ (in a nonobvious way). (See Remark A.48.)

Definition A.65 (Nullity). The *nullity* of a matrix $\mathbf{A} \in \mathbb{F}^{m \times n}$ is the dimension of the null space.

Theorem A.66 (Rank–nullity theorem). *Suppose* $\mathbf{A} \in \mathbb{F}^{m \times n}$. *Then, the sum of the rank and nullity of* $\mathbf{A}$ *is* n. □

Example A.67. We know that the matrix

$$\mathbf{A} = \begin{bmatrix} 1 & 2 & 3 \\ 2 & 4 & 6 \end{bmatrix}$$

is in $\mathbb{R}^{2 \times 3}$. We showed that its null space has a dimension of 2, and it is obvious that its rank is 1. We see at once that $1 + 2 = 3$.

A.9 Determinant

Remark A.68. The next definition uses concepts from abstract algebra covered in Definition 8.27. The reader who is not familiar with permutation groups should review this first.

Definition A.69 (Determinant). Let $\mathbf{A} \in \mathbb{F}^{n \times n}$. The *determinant* of $\mathbf{A}$ is

$$\det(\mathbf{A}) = \sum_{\sigma \in S_n} \operatorname{sgn}(\sigma) \prod_{i=1}^{n} \mathbf{A}_{i\sigma(i)}. \tag{A.17}$$

Here, $\sigma \in S_n$ represents a permutation over the set $\{1, \ldots, n\}$, and $\sigma(i)$ represents the value to which i is mapped under σ. The sign of the permutation $\operatorname{sgn}(\sigma)$ is given in Definition 8.38.

Example A.70. Consider an arbitrary 2×2 matrix:

$$\mathbf{A} = \begin{bmatrix} a & b \\ c & d \end{bmatrix}.$$

There are only two permutations in the set S_2: the identity permutation (which is even) and the transposition $(1, 2)$ (which is odd). Thus, we have

$$\det(\mathbf{A}) = \begin{vmatrix} a & b \\ c & d \end{vmatrix} = \mathbf{A}_{11}\mathbf{A}_{22} - \mathbf{A}_{12}\mathbf{A}_{21} = ad - bc.$$

This is the formula usually given in a course on matrix algebra.

Proposition A.71. *The determinant of any identity matrix is* 1. □

Remark A.72. Like many other definitions in mathematics, Definition A.69 can be useful for proving things but not very useful for computing determinants. Most linear algebra textbooks, such as Refs. [97,100,135,136], discuss formulas and algorithms for efficiently computing matrix determinants.

A.10 Eigenvalues and Eigenvectors

Definition A.73 (Algebraic closure). Let $\mathbb{F}$ be a field. The *algebraic closure* of $\mathbb{F}$, denoted by $\overline{\mathbb{F}}$, is an extension of $\mathbb{F}$ that is (i) also a field and (ii) has every possible root to any polynomial with coefficients drawn from $\mathbb{F}$.

Remark A.74. A field $\mathbb{F}$ is called *algebraically closed* if $\overline{\mathbb{F}} = \mathbb{F}$.

Theorem A.75. *The algebraic closure of* $\mathbb{R}$ *is* $\mathbb{C}$. *The field* $\mathbb{C}$ *is algebraically closed.* □

Definition A.76 (Eigenvalue and (right) eigenvector). Let $\mathbf{A} \in \mathbb{F}^{n\times n}$. An eigenvalue–eigenvector pair $(\lambda, \mathbf{x})$ is a scalar and $n \times 1$ a vector such that

$$\mathbf{A}\mathbf{x} = \lambda\mathbf{x} \tag{A.18}$$

and $\mathbf{x} \neq \mathbf{0}$. The eigenvalue may be drawn from $\overline{\mathbb{F}}$ and $\mathbf{x}$ from $\overline{\mathbb{F}}^n$.

Lemma A.77. *A value* $\lambda \in \mathbb{F}$ *is an eigenvalue of* $\mathbf{A} \in \mathbb{F}^{n\times n}$ *if and only if* $\lambda\mathbf{I}_n - \mathbf{A}$ *is not invertible.* □

Remark A.78. A *left eigenvector* is defined analogously with $\mathbf{x}^T\mathbf{A} = \lambda\mathbf{x}^T$, when $\mathbf{x}$ is considered a column vector. We deal exclusively with right eigenvectors, and hence, when we say "eigenvector," we mean a right eigenvector.

Definition A.79 (Characteristic polynomial). If $\mathbf{A} \in \mathbb{F}^{n\times n}$, then its *characteristic polynomial* is the degree-n polynomial

$$\det(\lambda\mathbf{I}_n - \mathbf{A}). \tag{A.19}$$

Theorem A.80. *A value λ is an eigenvalue of $\mathbf{A} \in \mathbb{F}^{n\times n}$ if and only if it satisfies the characteristic equation*

$$\det(\lambda \mathbf{I}_n - \mathbf{A}) = 0.$$

That is, λ is a root of the characteristic polynomial. □

Remark A.81. We now see why λ may be in $\overline{\mathbb{F}}$, rather than in $\mathbb{F}$. It is possible that the characteristic polynomial of a matrix does not have all (or any) of its roots in the field $\mathbb{F}$; the definition of algebraic closure ensures that all eigenvalues are contained in the the algebraic closure of $\mathbb{F}$.

Corollary A.82. *If $\mathbf{A} \in \mathbb{F}^{n\times n}$, then $\mathbf{A}$ and $\mathbf{A}^T$ share eigenvalues.* □

Example A.83. Consider the matrix

$$\mathbf{A} = \begin{bmatrix} 1 & 0 \\ 0 & 2 \end{bmatrix}.$$

The characteristic polynomial is computed as

$$\det(\lambda \mathbf{I}_n - \mathbf{A}) = \begin{vmatrix} \lambda - 1 & 0 \\ 0 & \lambda - 2 \end{vmatrix} = (\lambda - 1)(\lambda - 2) - 0 = 0.$$

Thus, the characteristic polynomial for this matrix is

$$\lambda^2 - 3\lambda + 2. \tag{A.20}$$

The roots of this polynomial are $\lambda_1 = 1$ and $\lambda_2 = 2$. Using these eigenvalues, we can compute eigenvectors as

$$\mathbf{x}_1 = \begin{bmatrix} 1 \\ 0 \end{bmatrix}, \tag{A.21}$$

$$\mathbf{x}_2 = \begin{bmatrix} 0 \\ 1 \end{bmatrix}, \tag{A.22}$$

and observe that

$$\mathbf{A}\mathbf{x}_1 = \begin{bmatrix} 1 & 0 \\ 0 & 2 \end{bmatrix} \begin{bmatrix} 1 \\ 0 \end{bmatrix} = 1 \begin{bmatrix} 1 \\ 0 \end{bmatrix} = \lambda_1 \mathbf{x}_1 \tag{A.23}$$

and

$$\mathbf{A}\mathbf{x}_2 = \begin{bmatrix} 1 & 0 \\ 0 & 2 \end{bmatrix} \begin{bmatrix} 0 \\ 1 \end{bmatrix} = 2 \begin{bmatrix} 0 \\ 1 \end{bmatrix} \lambda_2 \mathbf{x}_2, \tag{A.24}$$

as required. The computation of eigenvalues and eigenvectors is usually accomplished using a computer, for which several algorithms have been developed. Those interested readers should consult, for example, Ref. [100].

Remark A.84. You can use your calculator to return the eigenvalues and eigenvectors of a matrix, as well as several software packages, such as MATLAB™ and Mathematica™.

Remark A.85. It is important to remember that eigenvectors are unique *up to scale.* That is, if $\mathbf{A}$ is a square matrix and $(\lambda, \mathbf{x})$ is an eigenvalue–eigenvector pair for $\mathbf{A}$, then so is $(\lambda, \alpha\mathbf{x})$ for $\alpha \neq 0$. This is because

$$\mathbf{A}\mathbf{x} = \lambda\mathbf{x} \qquad \Longrightarrow \qquad \mathbf{A}(\alpha\mathbf{x}) = \lambda(\alpha\mathbf{x}). \tag{A.25}$$

Definition A.86 (Algebraic multiplicity of an eigenvalue). An eigenvalue has algebraic multiplicity greater than 1 if it is a *multiple root* of the characteristic polynomial. The *algebraic multiplicity* of the root is the *multiplicity* of the eigenvalue.

Example A.87. Consider the identity matrix $\mathbf{I}_2$. It has a characteristic polynomial of $(\lambda - 1)^2$, which has one multiple root 1 of multiplicity 2. However, this matrix does have two eigenvectors: $[1 \;\; 0]^T$ and $[0 \;\; 1]^T$.

Example A.88. Consider the matrix

$$\mathbf{A} = \begin{bmatrix} 1 & 5 \\ 2 & 4 \end{bmatrix}.$$

The characteristic polynomial is computed as

$$\begin{vmatrix} \lambda - 1 & -5 \\ -2 & \lambda - 4 \end{vmatrix} = (\lambda - 1)(\lambda - 4) - 10 = \lambda^2 - 5\lambda - 6.$$

Thus, there are two distinct eigenvalues: $\lambda = -1$ and $\lambda = 6$, which are the two roots of the characteristic polynomial. We can compute the two eigenvectors in turn. Consider $\lambda = -1$. We solve for

$$\begin{bmatrix} \lambda - 1 & -5 \\ -2 & \lambda - 4 \end{bmatrix} \begin{bmatrix} x_1 \\ x_2 \end{bmatrix} = \begin{bmatrix} -2 & -5 \\ -2 & -5 \end{bmatrix} \begin{bmatrix} x_1 \\ x_2 \end{bmatrix} = \begin{bmatrix} 0 \\ 0 \end{bmatrix}.$$

Thus,

$$-2x_1 - 5x_2 = 0.$$

We can set $x_2 = t$ as a free variable. Consequently, the solution is

$$\begin{bmatrix} x_1 \\ x_2 \end{bmatrix} = \begin{bmatrix} \frac{5}{2}t \\ t \end{bmatrix} = t \begin{bmatrix} \frac{5}{2} \\ 1 \end{bmatrix}.$$

Thus, any eigenvector of $\lambda = -1$ is a multiple of the vector $\langle 5/2, 1 \rangle$. For the eigenvalue $\lambda = -6$, we have

$$\begin{bmatrix} \lambda - 1 & -5 \\ -2 & \lambda - 4 \end{bmatrix} \begin{bmatrix} x_1 \\ x_2 \end{bmatrix} = \begin{bmatrix} 5 & -5 \\ -2 & 2 \end{bmatrix} \begin{bmatrix} x_1 \\ x_2 \end{bmatrix} = \begin{bmatrix} 0 \\ 0 \end{bmatrix}.$$

From this, we see that

$$-2x_1 + 2x_2 = 0,$$

or $x_1 = x_2$. Thus, setting $x_2 = t$, we have the solution

$$\begin{bmatrix} x_1 \\ x_2 \end{bmatrix} = \begin{bmatrix} t \\ t \end{bmatrix} = t \begin{bmatrix} 1 \\ 1 \end{bmatrix}.$$

Thus, any eigenvector of $\lambda = 6$ is a multiple of the vector $\langle 1, 1 \rangle$.

Theorem A.89. *Suppose that* $\mathbf{A} \in \mathbb{F}^{n \times n}$ *with eigenvalues* $\lambda_1, \ldots, \lambda_n$ *all distinct (i.e.,* $\lambda_i \neq \lambda_j$ *if* $i \neq j$*). Then, the corresponding eigenvectors* $\{\mathbf{v}_1, \ldots, \mathbf{v}_n\}$ *are linearly independent.* □

Definition A.90. Let $\mathbf{A} \in \mathbb{F}^{n \times n}$ with eigenvectors $\{\mathbf{v}_1, \ldots, \mathbf{v}_n\}$. Then, the vector space $\mathcal{E} = \text{span}(\{\mathbf{v}_1, \ldots, \mathbf{v}_n\})$ is called the *eigenspace* of $\mathbf{A}$. When the eigenvectors are linearly independent, they are an *eigenbasis* for the space they span.

Remark A.91. It is worth noting that if $\mathbf{v}_i$ is an eigenvector of $\mathbf{A}$, then $\text{span}(\mathbf{v}_i)$ is called the eigenspace associated with $\mathbf{v}_i$.

Definition A.92 (Geometric multiplicity). Each eigenvalue corresponds to a subset of eigenvectors. The *geometric multiplicity* is the dimension of the eigenspace to which it corresponds.

Example A.93. As we have seen, eigenvalue 1 for $\mathbf{I}_2$ corresponds to all of $\mathbb{R}^2$. Therefore, it has a geometric multiplicity of 2 as well as an algebraic multiplicity of 2.

Remark A.94. It is not the case that the geometric and algebraic multiplicities of an eigenvalue must be equal. In general, the algebraic multiplicity of an eigenvalue will be greater than or equal to its geometric multiplicity. However, if the algebraic multiplicity of an eigenvalue is 1, then this ensures that it will have a geometric multiplicity of 1.

Corollary A.95. *The eigenvectors of* $\mathbf{A} \in \mathbb{F}^{n\times n}$ *form a basis for* $\overline{\mathbb{F}}^n$ *when the eigenvalues of* $\mathbf{A}$ *are distinct.* □

Remark A.96. The following (and final) theorem is crucial to our understanding of algebraic graph theory. It is also a useful theorem in its own right and has generalizations in complex vector spaces $\mathbb{C}^n$ that have important consequences for quantum mechanics.

Theorem A.97 (Spectral theorem for real symmetric matrices). *Suppose* $\mathbf{A} \in \mathbb{R}^{n\times n}$ *is a real, symmetric matrix. Then,* $\mathbf{A}$ *has real eigenvalues, and the eigenvectors form an* <u>*orthonormal*</u> *basis for* $\mathbb{R}^n$. □

Remark A.98. The proof of the spectral theorem is given in almost every linear algebra textbook. Lang's presentation is straightforward [101]. The easy part is the proof that the eigenvalues are real. To see this, suppose that λ is a complex eigenvalue with a complex eigenvector $\mathbf{z}$. Then, $\mathbf{Az} = \lambda\mathbf{z}$. If $\lambda = a + bi$, then $\bar{\lambda} = a - bi$. Note that $\bar{\lambda}\lambda = a^2 + b^2 \in \mathbb{R}$. Furthermore, if λ and μ are two complex numbers, it's easy to show that $\overline{\lambda\mu} = \bar{\lambda}\bar{\mu}$. Since $\mathbf{A}$ is real (and symmetric), we can conclude that

$$\overline{\mathbf{Az}} = \mathbf{A}\bar{\mathbf{z}} = \bar{\lambda}\bar{\mathbf{z}}.$$

Now, $(\mathbf{A}\bar{\mathbf{z}})^T = \bar{\mathbf{z}}^T\mathbf{A} = \bar{\lambda}\bar{\mathbf{z}}^T$. We have

$$\begin{aligned} \mathbf{Az} = \lambda\mathbf{z} \qquad &\implies \qquad \bar{\mathbf{z}}^T\mathbf{Az} = \lambda\bar{\mathbf{z}}^T\mathbf{z}, \\ \bar{\mathbf{z}}^T\mathbf{A} = \bar{\lambda}\bar{\mathbf{z}}^T \qquad &\implies \qquad \bar{\mathbf{z}}^T\mathbf{Az} = \bar{\lambda}\bar{\mathbf{z}}^T\mathbf{z}. \end{aligned}$$

But then,

$$\lambda\bar{\mathbf{z}}^T\mathbf{z} = \bar{\lambda}\bar{\mathbf{z}}^T\mathbf{z}.$$

This implies that $\lambda = \bar{\lambda}$, which means λ *must* be real.

The rest of the proof, namely that the eigenvectors form an orthogonal basis, is the harder part and is well outside the scope of this appendix.

A.11 Exercises

Exercise A.1
Why is $\mathbb{Z}$ *not* a field under ordinary addition and multiplication? Is $\mathbb{Q}$, the set of rational numbers, a field under the usual addition and multiplication operations?

Exercise A.2
Show that $(\mathbb{R}^{2\times 1}, +, \mathbf{0})$ is a group.

Exercise A.3
Prove Proposition A.14.

Exercise A.4
Let $\mathbf{A}, \mathbf{B} \in \mathbb{R}^{m\times n}$. Use the definitions of matrix addition and transpose to prove that

$$(\mathbf{A} + \mathbf{B})^T = \mathbf{A}^T + \mathbf{B}^T. \tag{A.26}$$

[Hint: If $\mathbf{C} = \mathbf{A}+\mathbf{B}$, then $\mathbf{C}_{ij} = \mathbf{A}_{ij}+\mathbf{B}_{ij}$, the element in the (i, j)th position of matrix $\mathbf{C}$. This element moves to the (j, i)th position in the transpose. The (j, i)th position of $\mathbf{A}^T + \mathbf{B}^T$ is $\mathbf{A}^T_{ji} + \mathbf{B}^T_{ji}$, but $\mathbf{A}^T_{ji} = \mathbf{A}_{ij}$. Reason from this point onward.]

Exercise A.5
Let $\mathbf{A}, \mathbf{B} \in \mathbb{R}^{m\times n}$. Prove by example that $\mathbf{AB} \neq \mathbf{BA}$; that is, matrix multiplication is *not commutative*. [Hint: Almost any pair of matrices you pick (that can be multiplied) will not commute.]

Exercise A.6
Let $\mathbf{A} \in \mathbb{F}^{m\times n}$, and let $\mathbf{B} \in \mathbb{R}^{n\times p}$. Use the definitions of matrix multiplication and transpose to prove that

$$(\mathbf{AB})^T = \mathbf{B}^T\mathbf{A}^T \tag{A.27}$$

[Hint: Use similar reasoning to the hint in Question A.4. But this time, note that $\mathbf{C}_{ij} = \mathbf{A}_{i\cdot} \cdot \mathbf{B}_{\cdot j}$, which moves to the (j, i)th position. Now, figure out what is in the (j, i)th position of $\mathbf{B}^T\mathbf{A}^T$.]

Exercise A.7
Show that $(\mathbb{F}^{n\times n}, +, \mathbf{0})$ is a group, with $\mathbf{0}$ being the zero matrix.

Exercise A.8
Let $\mathbf{A} \in \mathbb{F}^{n\times n}$. Show that $\mathbf{A}\mathbf{I}_n = \mathbf{I}_n\mathbf{A} = \mathbf{A}$. Hence, $\mathbf{I}$ is an identify for the matrix multiplication operation on square matrices.

Exercise A.9
Prove that if $\mathbf{A}_1, \ldots, \mathbf{A}_n \in \mathbb{F}^{n\times n}$ are invertible, then $(\mathbf{A}_1, \ldots, \mathbf{A}_m)^{-1} = \mathbf{A}_m^{-1} \cdots \mathbf{A}_1^{-1}$ for $m \geq 1$.

Exercise A.10
Consider the vectors $\mathbf{v}_1 = \langle 0, 0\rangle$ and $\mathbf{v}_2 = \langle 1, 0\rangle$. Are these vectors linearly independent? Explain why or why not.

Exercise A.11
Show that the vectors

$$\mathbf{v}_1 = \begin{bmatrix} 1 \\ 2 \\ 3 \end{bmatrix}, \quad \mathbf{v}_2 = \begin{bmatrix} 4 \\ 5 \\ 6 \end{bmatrix}, \quad \text{and} \quad \mathbf{v}_3 = \begin{bmatrix} 7 \\ 8 \\ 9 \end{bmatrix}$$

are *not* linearly independent. [Hint: Following the examples, create a system of equations and show that there is a solution not equal to $\mathbf{0}$.]

Exercise A.12
Why are the vectors

$$\mathbf{v}_1 = \begin{bmatrix} 1 \\ 2 \\ 3 \end{bmatrix}, \quad \mathbf{v}_2 = \begin{bmatrix} 4 \\ 5 \\ 6 \end{bmatrix}, \quad \text{and} \quad \mathbf{v}_3 = \begin{bmatrix} 7 \\ 8 \\ 9 \end{bmatrix}$$

not a basis for $\mathbb{R}^3$.

Exercise A.13
Prove Proposition A.63.

Exercise A.14
Prove the Proposition A.71.

Exercise A.15
Find the eigenvalues and eigenvectors of the matrix

$$\mathbf{A} = \begin{bmatrix} 1 & 2 \\ 2 & 1 \end{bmatrix}.$$

Exercise A.16
Show that every vector in $\mathbb{F}^2$ is an eigenvector of $\mathbf{I}_2$.

Exercise A.17
Prove Corollary A.82.

Exercise A.18
Prove Corollary A.95.

Appendix B

A Brief Introduction to Probability Theory

Remark B.1 (Chapter goals). This appendix provides a brief (yet theoretical) introduction to probability theory for students who have no background in probability. Proofs are omitted for brevity.

B.1 Probability

Remark B.2. The proper study of probability theory requires a heavy dose of measure theory, which is well beyond the scope of a course in graph theory. This appendix is meant to provide a somewhat intuitive introduction to probability theory with a small amount of mathematical rigor added. This is more than sufficient to understand the probability calculations that occur when we discuss Markov chains in Chapter 10. Most of the definitions in this chapter are motivated by examples from games. We'll begin with an example.

Example B.3. Suppose you have made it to the very final stage of *Deal or No Deal.* Two suitcases with money remain in play, one contains $0.01 while the other contains $1,000,000. The banker has offered you a payoff of $499,999. Do you accept the banker's safe offer or do you risk it all to try for $1,000,000. Suppose the banker offers you $100,000, what about $500,000 or $10,000?

Definition B.4 (Outcome). Let Ω be a finite set of elements describing the outcome of a chance event (a coin toss, a roll of the dice, etc.). We call Ω the *sample space*. Each element of Ω is called an *outcome*.

Example B.5. Congratulations! You have made it to the very final stage of *Deal or No Deal.* Two suitcases with money remain in play, one contains \$0.01 while the other contains \$1,000,000. The banker has offered you a payoff of \$499,999. Do you accept the banker's safe offer or do you risk it all to try for \$1,000,000. Suppose the banker offers you \$100,000 what about \$500,000 or \$10,000? To even begin to reason about this situation, we note that the world as we care about is purely the position of \$1,000,000 and \$0.01 within the suitcases. In this case, Ω consists of two possible outcomes: \$1,000,000 is in suitcase number 1 (while \$0.01 is in suitcase number 2) or \$1,000,000 is in suitcase number 2 (while \$0.01 is in suitcase number 1).

Formally, let us refer to the first outcome as A and the second outcome as B. Then, $\Omega = \{A, B\}$.

Definition B.6 (Event). If Ω is a sample space, then an event is any subset of Ω.

Example B.7. Clearly, the sample space in Example B.3 consists of precisely four events: $\emptyset$ (the empty event), $\{A\}$, $\{B\}$, and $\{A, B\} = \Omega$. These four sets represent all possible subsets of the set $\Omega = \{A, B\}$.

Definition B.8 (Union). If $E, F \subseteq \Omega$ are both events, then $E \cup F$ is the *union* of the sets E and F and consists of all outcomes in either E or F. Event $E \cup F$ occurs if either event E or event F occurs.

Example B.9. Consider the role of a fair six-sided dice. The outcomes are 1, ..., 6. If $E = \{1, 3\}$ and $F = \{2, 4\}$, then $E \cup F = \{1, 2, 3, 4\}$ and will occur as long as we don't roll a 5 or 6.

Definition B.10 (Intersection). If $E, F \subseteq \Omega$ are both events, then $E \cap F$ is the *intersection* of the sets E and F and consists of all outcomes in both E and F. Event $E \cap F$ occurs if both even E and event F occur.

Example B.11. Again, consider the role of a fair six-sided dice. The outcomes are 1,. . . ,6. If $E = \{1, 2\}$ and $F = \{2, 4\}$, then $E \cap F = \{2\}$ and will occur only if we roll a 2.

Definition B.12 (Mutual exclusivity). Two events $E, F \subseteq \Omega$ are said to be *mutually exclusive* if and only if $E \cap F = \emptyset$.

Definition B.13 (Discrete probability distribution (function)). Given a discrete sample space Ω, let $\mathcal{F}$ be the set of all events on Ω. A *discrete probability function* is a mapping from $P : \mathcal{F} \to [0, 1]$ with the following properties:

(1) $P(\Omega) = 1$.
(2) If $E, F \in \mathcal{F}$ and $E \cap F = \emptyset$, then $P(E \cup F) = P(E) + P(F)$.

Remark B.14 (Power set). In this definition, we talk about the set $\mathcal{F}$ as the set of all events over a set of outcomes Ω. This is an example of the *power set*: the set of all subsets of a set. We sometimes denote this set as 2^{Ω}. Thus, if Ω is a set, then 2^{Ω} is the power set of Ω or the set of all subsets of Ω. (See Definition 5.54.)

Remark B.15. Definition B.13 is surprisingly technical and probably does not conform to your ordinary sense of what probability is. It's best not to think of probability in this very formal way and instead to think that a probability function assigns a number to an outcome (or event) that tells you the chances of it occurring. Put more simply, suppose we could run an experiment where the result of that experiment will be an outcome in Ω. The function P simply tells us the proportion of times we will observe an event $E \subset \Omega$ if we ran this experiment an exceedingly large number of times.

Example B.16. Suppose we could play the *Deal or No Deal* example over and over again and observe where the money ends up. A smart game show would mix the money up so that approximately one half of the time we observe \$1,000,000 in suitcase 1 and the other half the time we observe this money in suitcase 2.

A probability distribution formalizes this notion and might assign 1/2 to event $\{A\}$ and 1/2 to event $\{B\}$. However, to obtain a true probability distribution, we must also assign probabilities to $\emptyset$ and $\{A, B\}$. In the former case, we know that something must happen!

Therefore, we can assign 0 to event $\emptyset$. In the latter case, we know that for certain that either outcome A or outcome B must occur; so, in this case, we assign a value of 1.

Example B.17. In a fair six-sided dice, the probability of rolling any value is 1/6. Formally, $\Omega = \{1, 2, \ldots, 6\}$ any role yields is an event with only one element, $\{\omega\}$, where ω is some value in Ω. If we consider the event $E = \{1, 2, 3\}$, then $P(E)$ gives us the probability that we will roll a 1, 2, or 3. Since $\{1\}$, $\{2\}$, and $\{3\}$, are disjoint sets and $\{1, 2, 3\} = \{1\} \cup \{2\} \cup \{3\}$, we know that

$$P(E) = \frac{1}{6} + \frac{1}{6} + \frac{1}{6} = \frac{1}{2}$$

Definition B.18 (Discrete probability space). The triple $(\Omega, \mathcal{F}, P)$ is called a *discrete probability space* over Ω.

Lemma B.19. *Let $(\Omega, \mathcal{F}, P)$ be a discrete probability space. Then, $P(\emptyset) = 0$.* □

Lemma B.20. *Let $(\Omega, \mathcal{F}, P)$ be a discrete probability space, and let $E, F \in \mathcal{F}$. Then,*

$$P(E \cup F) = P(E) + P(F) - P(E \cap F). \tag{B.1}$$

□

Definition B.21 (Set complement). Let Ω be a set of outcomes. Let $E \subseteq \Omega$, and define E^c to be the set of elements of Ω *not* in E. This is called the *complement* of E in Ω.

Lemma B.22. *Let $(\Omega, \mathcal{F}, P)$ be a discrete probability space, and let $E, F \in \mathcal{F}$. Then,*

$$P(E) = P(E \cap F) + P(E \cap F^c). \tag{B.2}$$

□

Theorem B.23. *Let $(\Omega, \mathcal{F}, P)$ be a discrete probability space and let $E \in \mathcal{F}$. Let $F_1, \ldots, F_n$ be any pairwise-disjoint collection of sets that* *partition* Ω*. That is, assume*

$$\Omega = \bigcup_{i=1}^{n} F_i \tag{B.3}$$

and $F_i \cap F_j = \emptyset$ *if* $i \neq j$*. Then,*

$$P(E) = \sum_{i=1}^{n} P(E \cap F_i). \tag{B.4}$$

□

Example B.24. Welcome to Vegas! We're playing craps. In craps, we roll two dice, and winning combinations are determined by the sum of the values on the dice. An ideal first craps roll is 7. The sample space Ω, in which we are interested, has 36 elements, one each for the possible values the dice will show (the related set of sums can be easily obtained).

Suppose that the dice are colored blue and red (so they can be distinguished), and let's call the blue die number one and the red die number two. Let's suppose we are interested in the event that we roll a 1 on die number one and that the pair of values obtained sums to 7. There is only one way this can occur, namely, we roll a 1 on die number one and a 6 on die number two. Thus, the probability of this occurring is 1/36. In this case, event E is the event that we roll a 7 in our craps game and event F_1 is the event that die number one shows a 1. We could also consider event F_2 that die number one shows a 2. By similar reasoning, we know that the probability of both E and F_2 occurring is 1/36. In fact, if F_i is the event that one of the dice shows a value of i ($i = 1, \ldots, 6$), then we know that

$$P(E \cap F_i) = \frac{1}{36}.$$

Clearly, the events F_i ($i = 1, \ldots, 6$) are pairwise disjoint (you can't have both a 1 and a 2 on the same die). Furthermore, $\Omega = F_1 \cup F_2 \cup \cdots \cup F_6$. (After all, some number has to appear on die number one!) Thus, we can compute

$$P(E) = \sum_{i=1}^{6} P(E \cap F_i) = \frac{6}{36} = 16.$$

B.2 Random Variables and Expected Values

Remark B.25. The concept of a random variable can be made extremely mathematically specific. A good intuitive understanding of a random variable is a variable X whose value is not known *a priori* and which is determined according to some probability distribution P that is a part of a probability space $(\Omega, \mathcal{F}, P)$.

Example B.26. Suppose that we consider flipping a fair coin. Then, the probability of seeing *heads* (or *tails*) should be 1/2. If we let X be a random variable that provides the outcome of the flip, then it will take on values *heads* or *tails*, and it will take each value exactly 50% of the time.

Remark B.27. The problem with allowing a random variable to take on arbitrary values (such as *heads* or *tails*) is that it makes it difficult to use random variables in formulae involving numbers. There is a *very* technical definition of random variable that arises in formal probability theory. However, it is well beyond the scope of what we need. We can, however, get a flavor for this definition in the following restricted form.

Definition B.28 (Random variable). Let $(\Omega, \mathcal{F}, P)$ be a discrete probability space. Let $D \subseteq \mathbb{R}$ be a finite discrete subset of real numbers. A random variable X is a function that maps each element of Ω to an element of D. Formally, $X : \Omega \to D$.

Remark B.29. Clearly, if $S \subseteq D$, then $X^{-1}(S) = \{\omega \in \Omega | X(\omega) \in S\} \in \mathcal{F}$. We can think of the probability of X taking on a value in $S \subseteq D$ is precisely $P(X^{-1}(S))$.

Using this observation, if $(\Omega, \mathcal{F}, P)$ is a discrete probability distribution function and $X : \Omega \to D$ is a random variable and $x \in D$, then let $P(x) = P(X^{-1}(\{x\})$. That is, the probability of X taking a value of x is the probability of the element in Ω corresponding to x.

Example B.30. Consider our coin-flipping random variable. Instead of having X take values *heads* or *tails*, we can instead let X take on values 1 if the coin comes up *heads* and 0 if the coin comes up *tails*. Thus, if $\Omega = \{\textit{heads}, \textit{tails}\}$, then $X(\textit{heads}) = 1$ and $X(\textit{tails}) = 0$.

Example B.31. When Ω (in probability space $(\Omega, \mathcal{F}, P)$) is already a subset of $\mathbb{R}$, then defining random variables is very easy. The random variable can just be the obvious mapping from Ω into itself. For example, if we consider rolling a fair die, then $\Omega = \{1, \ldots, 6\}$ and any random variable defined on $(\Omega, \mathcal{F}, P)$ will take on values $1, \ldots, 6$.

Definition B.32. Let $(\Omega, \mathcal{F}, P)$ be a discrete probability distribution, and let $X : \Omega \to D$ be a random variable. Then, the *expected value* of X is

$$\mathbb{E}(X) = \sum_{x \in D} x P(x). \tag{B.5}$$

Example B.33. Let's play a die-rolling game. You put up your own money. Even numbers lose \$10 times the number rolled, while odd numbers win \$12 times the number rolled. What is the expected amount of money you'll win in this game?

Let $\Omega = \{1, \ldots, 6\}$. Then, $D = \{12, -20, 36, -40, 60, -60\}$: these are the dollar values you will win for various rolls of the dice. Then, the expected value of X is

$$\begin{aligned} \mathbb{E}(X) = 12 \left(\frac{1}{6}\right) + (-20) \left(\frac{1}{6}\right) + 36 \left(\frac{1}{6}\right) \\ + (-40) \left(\frac{1}{6}\right) + 60 \left(\frac{1}{6}\right) + (-60) \left(\frac{1}{6}\right) = -2. \end{aligned} \tag{B.6}$$

Would you still want to play this game considering the expected payoff is $-\$2$?

B.3 Conditional Probability

Remark B.34. Suppose we are given a discrete probability space $(\Omega, \mathcal{F}, P)$ and we are told that an event E has occurred. We now wish to compute the probability that some other event F has occurred. This value is called the conditional probability of event F given event E and is written as $P(F|E)$.

Example B.35. Suppose we roll a fair six-sided die twice. The sample space in this case is the set $\Omega = \{(x, y) | x = 1, \ldots, 6, \ y = 1, \ldots, 6\}$. Suppose I roll a 2 on the first try. I want to know what the probability of rolling a combined score of 8 is. That is, given that I've rolled a 2, I wish to determine the conditional probability of rolling a 6.

Since the die is fair, the probability of rolling any pair of values $(x, y) \in \Omega$ is equally likely. There are 36 elements in Ω, so each is assigned a probability of 1/36. That is, $(\Omega, \mathcal{F}, P)$ is defined so that $P((x, y)) = 1/36$ for each $(x, y) \in \Omega$.

Let E be the event that we roll a 2 on the first try. We wish to assign a new set of probabilities to the elements of Ω to reflect this information. We know that our final outcome must have the form $(2, y)$, where $y \in \{1, \ldots, 6\}$. In essence, E becomes our new sample space. Furthermore, we know that each of these outcomes is equally likely because the die is fair. Thus, we may assign $P((2, y)|E) = 1/6$ for each $y \in \{1, \ldots, 6\}$ and $P((x, y)|E) = 0$ just in case $x \neq 2$, so $(x, y) \notin E$. This last definition occurs because we know that we've already observed a 2 on the first roll, so it's impossible to see another first number not equal to 2.

At last, we can answer the question we originally posed. The only way to obtain a sum equal to 8 is to roll a six on the second attempt. Thus, the probability of rolling a combined score of 8 given a 2 on the first roll is 1/6.

Lemma B.36. *Let $(\Omega, \mathcal{F}, P)$ be a discrete probability space and suppose that event $E \subseteq \Omega$. Then, $(E, \mathcal{F}_E, P_E)$ is a discrete probability space when*

$$P_E(F) = \frac{P(F)}{P(E)} \tag{B.7}$$

for all $F \subseteq E$ and $P_E(\omega) = 0$ for any $\omega \notin E$. □

Remark B.37. The previous lemma gives us a direct way to construct $P(F|E)$ for arbitrary $F \subseteq \Omega$. Clearly, if $F \subseteq E$, then

$$P(F|E) = P_E(F) = \frac{P(F)}{P(E)}.$$

Now, suppose that F is not a subset of E but that $F \cap E \neq \emptyset$. Then, clearly, the only possible events that can occur in F, given that E has occurred, are the ones that are also in E. Thus, $P_E(F) = P_E(E \cap F)$.

More to the point, we have

$$P(F|E) = P_E(F \cap E) = \frac{P(F \cap E)}{P(E)}. \tag{B.8}$$

Definition B.38 (Conditional probability). Given a discrete probability space $(\Omega, \mathcal{F}, P)$ and an event $E \in \mathcal{F}$, the conditional probability of event $F \in \mathcal{F}$ given event E is

$$P(F|E) = \frac{P(F \cap E)}{P(E)}. \tag{B.9}$$

Example B.39 (Simple blackjack). Blackjack is a game in which decisions can be made entirely based on conditional probabilities. The chances of a card appearing are based entirely on whether or not you have seen that card already since cards are discarded as the dealer works her way through the deck.

Consider a simple game of blackjack played with only the cards A, 2, 3, 4, 5, 6, 7, 8, 9, 10, J, Q, and K. In this game, the dealer deals two cards to the player and two to herself. The objective is to obtain a score as close to 21 as possible without going over. Face cards are worth 10, A is worth 1 or 11, and all other cards are worth their respective face value. We'll assume that the dealer must hit (take a new card) on 16 and below and will stand on 17 and above.

The complete sample space in this case is very complex; it consists of all possible valid hands that could be dealt over the course of a standard play of the game. We can however consider a simplified sample space of hands after the initial deal. In this case, the sample space has the form

$$\Omega = \{(\langle x, y\rangle, \langle s, t\rangle)\}.$$

Here, x, y, s, and t are cards without repeats. The total size of the sample space is

$$13 \times 12 \times 11 \times 10 = 17{,}160.$$

This can be seen by noting that the player can receive any of the 13 cards as the first card and any of the remaining 12 cards for the second card. The dealer then receives one of the 11 remaining cards and then one of the 11 remaining cards.

Let's suppose that the player is dealt 10 and 6 for a score of 16, while the dealer receives a 4 and 5 for a total of 9. If we suppose that the player decides to hit, then the large sample space (Ω) becomes

$$\Omega = \{(\langle x, y, z\rangle, \langle s, t\rangle)\},$$

which has a size of

$$13 \times 12 \times 11 \times 10 \times 9 = 154{,}440,$$

while the event is

$$E = \{(\langle 10, 6, z\rangle, \langle 4, 5\rangle)\}.$$

There are nine possible values for z, and thus, $P(E) = 9/154{,}440$.

Let us now consider the probability of busting on our first hit. This is event F and is given as

$$F = \{(\langle x, y, z\rangle, \langle s, t\rangle) : x + y + z > 21\}.$$

(Here, we take some liberty by assuming that we can add card values like digits.)

The set F is very complex, but we can see immediately that

$$E \cap F = \{(\langle 10, 6, z\rangle, \langle 4, 5\rangle) : z \in \{7, 8, 9, J, Q, K\}\}$$

because these are the hands that will cause us to bust. Thus, we can easily compute

$$P(F|E) = \frac{P(E \cap F)}{P(E)} = \frac{6/154{,}440}{9/154{,}440} = \frac{6}{9} = \frac{2}{3}. \qquad \text{(B.10)}$$

Thus, the probability of not busting given the hand we have drawn must be 1/3. We can see at once that our odds when taking a hit are not very good. Depending on the probabilities associated with the dealer busting, it may be smarter for us to not take a hit and see what happens to the dealer; however, in order to be sure we'd have to work out the chances of the dealer busting (since we know she will continue to hit until she busts or exceeds our value of 16). This computation is quite tedious, so we will not include it here.

Remark B.40. The complexity associated with blackjack makes knowing exact probabilities difficult, if not impossible. Thus, most card-counting strategies use heuristics to attempt to understand approximately what the probabilities are for winning given the history of observed hands. To do this, simple numeric values are assigned to cards: generally a +1 to cards with low values (2, 3, 4, etc.), a 0 to cards with mid-range values (7, 8, and 9), and negative values for face cards (10, J, Q, and K). As the count gets *high*, there are more face cards in the deck, and thus, the chances of the dealer busting or the player drawing blackjack increase. If the count is low, there are fewer face cards in the deck, and the chance of the dealer drawing a sufficient number of cards without busting is higher. Thus, players favor tables with high counts.

Definition B.41 (Independence). Let $(\Omega, \mathcal{F}, P)$ be a discrete probability space. Two events $E, F \in \mathcal{F}$ are called *independent* if $P(E|F) = P(E)$ and $P(F|E) = P(F)$.

Theorem B.42. *Let $(\Omega, \mathcal{F}, P)$ be a discrete probability space. If $E, F \in \mathcal{F}$ are independent events, then $P(E \cap F) = P(E)P(F)$.* □

Example B.43. Consider rolling a fair die twice in a row. Let Ω be the sample space of pairs of die results that will occur. Thus, $\Omega = \{(x, y)|x = 1, \ldots, 6,\ y = 1, \ldots, 6\}$. Let E be the event that says we obtain a 6 on the first roll. Then, $E = \{(6, y) : y = 1, \ldots, 6\}$, and let F be the event that says we obtain a 6 on the second roll. Then, $F = \{(x, 6) : x = 1, \ldots, 6\}$. Obviously, these two events are independent. The first roll *cannot* affect the outcome of the second roll, thus $P(F|E) = P(F)$. We know that $P(E) = P(F) = 1/6$. That is, there is a one in six chance of observing a 6. Thus, the chance of rolling double sixes in two rolls is precisely the probability of both events E and F occurring. Using our result on independent events, we can see that $P(E \cap F) = P(E)P(F) = (1/6)^2 = 1/36$, just as we expect it to be.

Remark B.44. The previous result will help in understanding Theorem 10.16.

B.4 Exercises

Exercise B.1
A fair four-sided die is rolled. Assume the sample space of interest to be the number appearing on the die and the numbers run from 1 to 4. Identify the space Ω precisely and all the possible outcomes and events within the space. What is the (logical) fair probability distribution in this case. [Hint: See Example B.17.]

Exercise B.2
Prove the following: Let $E \subseteq \Omega$, and define E^c to be the set of elements of Ω *not* in E (this is called the complement of E). Suppose $(\Omega, \mathcal{F}, P)$ is a discrete probability space. Show that $P(E^c) = 1 - P(E)$.

Exercise B.3
Prove Lemma B.22. [Hint: Show that $E \cap F$ and $E \cap F^c$ are mutually exclusive events. Then, show that $E = (E \cap F) \cup (E \cap F^c)$.]

Exercise B.4
Suppose that I change the definition of F_i in Example B.24 to read: value i appears on either die, while keeping the definition of event E the same. Do we still have

$$P(E) = \sum_{i=1}^{6} P(E \cap F_i)?$$

If so, show the computation. If not, explain why.

Exercise B.5
Use Definition B.38 to compute the probability of obtaining a sum of 8 in two rolls of a die given that in the first roll, a 1 or 2 appears. [Hint: The space of outcomes is still $\Omega = \{(x, y)|x = 1, \ldots, 6,\ y = 1, \ldots, 6\}$. First, identify the event E within this space. How many elements within this set will enable you to obtain an 8 in two rolls? This is the set $E \cap F$. What is the probability of $E \cap F$? What is the probability of E? Use the formula in Definition B.38. It might help to write out the space Ω.]

References

[1] U. C. Merzbach and C. B. Boyer, *A History of Mathematics*. John Wiley & Sons Ltd., Hoboken, NJ, USA (2011).

[2] Mathematical Association of America, The Euler Archive (2022), http://eulerarchive.maa.org/correspondence/correspondents/Marino ni.html.

[3] S. Choudum, A simple proof of the Erdos-Gallai theorem on graph sequences, *Bulletin of the Australian Mathematical Society* **33**(1), pp. 67–70 (1986), DOI: 10.1017/S0004972700002872.

[4] D. Heiligman, *The Boy Who Loved Math: The Improbable Life of Paul Erdos*. Roaring Brook Press, New York, NY, USA (2013).

[5] Mathematical Reviews (MathSciNet), Mathematical Reviews: Paul Erös (2021), https://mathscinet.ams.org/mathscinet/search/authors.html?authorName=Erdos+Paul.

[6] L. Lovász and M. D. Plummer, *Matching Theory*, Vol. 367. American Mathematical Society, Providence, RI, USA (2009).

[7] V. Havel, A remark on the existence of finite graphs, *Casopis Pro Pestovani Matematiky* **80**, pp. 477–480 (1955).

[8] S. L. Hakimi, On realizability of a set of integers as degrees of the vertices of a linear graph. I, *Journal of the Society for Industrial and Applied Mathematics* **10**(3), pp. 496–506 (1962).

[9] M. Newman, *Networks: An Introduction*. Oxford University Press, Oxford, England (2018).

[10] A. Barabási and R. Albert, Emergence of scaling in random networks, *Science* **286**(509), pp. 509–512 (1999).

[11] L. A. Adamic and B. A. Huberman, Power-law distribution of the world wide web, *Science* **287**(5461), pp. 2115–2115 (2000).

[12] M. Molloy and B. Reed, A critical point for random graphs with a given degree sequence, *Random Structures Algorithms* **6**, pp. 161–179 (1995).

[13] W. Aiello, F. Chung and L. Lu, A random graph model for power law graphs, *Experimental Mathematics* **10**(1), pp. 53–66 (2001).

[14] B. Bollobás and O. Riordan, Robustness and vulnerability of scalefree random graphs, *Internet Mathematics* **1**, pp. 1–35 (2003).

[15] L. Lu, The diameter of random massive graphs, in *Proceedings of the Twelfth Annual ACM-SIAM Symposium on Discrete Algorithms*, pp. 912–921 (2001).

[16] J. O'rourke *et al.*, *Art Gallery Theorems and Algorithms*, Vol. 57. Oxford University Press, Oxford (1987).

[17] D. Knoke and S. Yang, *Social Network Analysis*, Vol. 154. Quantitative Applications in the Social Sciences. SAGE Publications, Thousand Oaks, CA, USA (2008).

[18] P. J. Carrington, J. Scott and S. Wasserman, *Models and Methods in Social Network Analysis.* Cambridge University Press, Cambridge, England (2005).

[19] J. Gross and J. Yellen, *Graph Theory and Its Applications*, 2nd edn. CRC Press, Boca Raton, FL, USA (2005).

[20] B. Bollobás, *Extremal Graph Theory.* Dover Press, Mineola, NY, USA (2004).

[21] S. T. Thornton and J. B. Marion, *Classical Dynamics of Particles and Systems*, 5th edn. Cengage Learning, Boston, MA, USA (2003).

[22] N. L. Biggs, T. P. Kirkman, mathematician, *Bulletin of the London Mathematical Society* **13**(2), pp. 97–120 (1981).

[23] J. J. Watkins, Chapter 2: Knight's tours, in *Across the Board: The Mathematics of Chessboard Problems.* Princeton University Press, Princeton, NJ, USA (2004).

[24] R. Diestel, *Graph Theory*, 4th edn. Graduate Texts in Mathematics. Springer, Berlin/Heidelberg, Germany (2010).

[25] W. W. Zachary, An information flow model for conflict and fission in small groups, *Journal of Anthropological Research* **33**(4), pp. 452–473 (1977).

[26] L. Katz, A new status index derived from sociometric analysis, *Psychometrika* **18**(1), pp. 39–43 (1953).

[27] S. Brin and L. Page, The anatomy of a large-scale hypertextual web search engine, in *Seventh International World-Wide Web Conference (WWW 1998)* (1998).

[28] T. Cormen, C. Leiserson, R. Rivest and C. Stein, *Introduction to Algorithms*, 2nd edn. The MIT Press, Cambridge, MA, USA (2001).

[29] M. S. Bazaraa, J. J. Jarvis and H. D. Sherali, *Linear Programming and Network Flows*. Wiley-Interscience, Hoboken, NJ, USA (2004), ISBN 0471485993.

[30] R. D. Luce and H. Raiffa, *Games and Decisions: Introduction and Critical Survey*. Dover Press, Mineola, NY, USA (1989).

[31] R. B. Myerson, *Game Theory: Analysis of Conflict*. Harvard University Press, Cambridge, MA, USA (2001).

[32] T. K. Moon, *Error Correction Coding: Mathematical Methods and Algorithms*. John Wiley & Sons, Hoboken, NJ, USA (2020).

[33] G. Rozenberg and J. Engelfriet, Elementary net systems, in *Advanced Course on Petri Nets*. Springer, Berlin/Heidelberg, Germany, pp. 12–121 (1996).

[34] S. Wasserman, K. Faust *et al.*, *Social Network Analysis: Methods and Applications*, Cambridge University Press, Cambridge, England (1994).

[35] D. Knuth, *Sorting and Searching, The Art of Computer Programming*, Vol. 3. Addison-Wesley, Boston, MA, USA (1998).

[36] P. A. Pevzner, H. Tang and M. S. Waterman, An Eulerian path approach to DNA fragment assembly, *Proceedings of the National Academy of Sciences* **98**(17), pp. 9748–9753 (2001).

[37] K. Roy, Optimum gate ordering of CMOS logic gates using Euler path approach: Some insights and explanations, *Journal of Computing and Information Technology* **15**(1), pp. 85–92 (2007).

[38] C. B. Hurley and R. Oldford, Pairwise display of high-dimensional information via Eulerian tours and hamiltonian decompositions, *Journal of Computational and Graphical Statistics* **19**(4), pp. 861–886 (2010).

[39] C. Godsil and G. Royle, *Algebraic Graph Theory*. Springer, Berlin/Heidelberg, Germany (2001).

[40] M.-K. Kwan, Graphic programming using odd or even points, *Acta Mathematica Sinica* **10**, pp. 263–266 (1962).

[41] J. Hopcroft and J. D. Ullman, *Introduction to Automata Theory, Languages and Computation*. Addison-Wesley, Reading, MA (1979).

[42] R. C. Prim, Shortest connection networks and some generalizations, *Bell System Technical Journal* **36** (1957), pp. 1389–1401.

[43] J. B. Kruskal, On the shortest spanning subtree of a graph and the traveling salesman problem, *Proceedings of AMS* **7**(1) (1956), pp. 48–50.

[44] E. W. Dijkstra, A note on two problems in connexion with graphs, *Numerische Mathematik* **1**, pp. 269–271, (1959).

[45] R. W. Floyd, Algorithm 97: Shortest path, *Communications of the ACM* **5**(6), p. 345 (1962).

[46] R. Bellman, *Dynamic Programming.* Princeton University Press, Princeton, NJ, USA (1957).

[47] J. G. Oxley, *Matroid Theory,* 2nd edn. Oxford University Press, Oxford, England (2011).

[48] R. C. Prim, Shortest connection networks and some generalizations, *The Bell System Technical Journal* **36**(6), pp. 1389–1401 (1957).

[49] V. Jarník, O jistém problému minimálním [about a certain minimal problem], *Práce Moravské Přírodovědecké Společnosti* **6**(4), pp. 57–63 (1930).

[50] J. B. Kruskal, On the shortest spanning subtree of a graph and the traveling salesman problem, *Proceedings of the American Mathematical Society* **7**(1), pp. 48–50 (1956).

[51] H. Loberman and A. Weinberger, Formal procedures for connecting terminals with a minimum total wire length, *Journal of the ACM (JACM)* **4**(4), pp. 428–437 (1957).

[52] E. W. Dijkstra *et al.*, A note on two problems in connexion with graphs, *Numerische Mathematik* **1**(1), pp. 269–271 (1959).

[53] Association for Computing Machines, EDSGER WYBE DIJKSTRA, https://amturing.acm.org/award_winners/dijkstra_1053701.cfm (Last Accessed, June 2022).

[54] B. Roy, Transitivité et connexité, *Comptes Rendus Hebdomadaires Des Seances De L Academie Des Sciences* **249**(2), pp. 216–218 (1959).

[55] R. W. Floyd, Algorithm 97: Shortest path, *Communications of the ACM* **5**(6), p. 345 (1962).

[56] S. Warshall, A theorem on boolean matrices, *Journal of the ACM (JACM)* **9**(1), pp. 11–12 (1962).

[57] H. Whitney, On the abstract properties of linear dependence, in *Classic Papers in Combinatorics.* Springer, Berlin/Heidelberg, Germany pp. 63–87 (2009).

[58] H. Nishimura and S. Kuroda, *A Lost Mathematician, Takeo Nakasawa: The Forgotten Father of Matroid Theory.* Springer, Berlin-Heidelberg, Germany (2009).

[59] B. Korte and J. Vygen, *Combinatorial Optimization.* Springer-Verlag, Berlin-Heidelberg, Germany (2008).

[60] U. Zwick, The smallest networks on which the Ford-Fulkerson maximum flow procedure may fail to terminate, *Theoretical Computer Science* **148**(1), pp. 165–170 (1995).

[61] B. L. Schwartz, Possible winners in partially completed tournaments, *SIAM Review* **8**(3), pp. 302–308 (1966).

[62] L. R. Ford and D. R. Fulkerson, Flows in networks, in *Flows in Networks.* Princeton University Press, Princeton, NJ, USA (2015).

[63] L. R. Ford and D. R. Fulkerson, Maximal flow through a network, *Canadian Journal of Mathematics* **8**, pp. 399–404 (1956).

[64] J. Edmonds and R. M. Karp, Theoretical improvements in algorithmic efficiency for network flow problems, *Journal of the ACM (JACM)* **19**(2), pp. 248–264 (1972).

[65] Y. Dinitz, Algorithm for solution of a problem of maximum flow in a network with power estimation, *Soviet Mathematics Doklady* **11**(5), pp. 1277–1280 (1970) [*Doklady Akademii Nauk SSSR* **11**, pp. 1277–1280 (1970)].

[66] J. B. Orlin, Max flows in o (nm) time, or better, in *Proceedings of the Forty-Fifth Annual ACM Symposium on Theory of Computing*, pp. 765–774 (2013).

[67] L. Chen, R. Kyng, Y. P. Liu, R. Peng, M. P. Gutenberg and S. Sachdeva, Maximum flow and minimum-cost flow in almost-linear time, *arXiv preprint arXiv:2203.00671* (2022).

[68] G. Chartrand and H. Kronk, Randomly traceable graphs, *SIAM Journal on Applied Mathematics* **16**(4), pp. 696–700 (1968).

[69] A. Bondy and U. S. R. Murty, *Graph Theory*, 3rd edn. Graduate Texts in Mathematics. Springer, Berlin/Heidelberg, Germany (2008).

[70] S. Simpson, Mathematical logic (2005), https://www.personal.psu.edu/t20/notes/logic.pdf.

[71] R. M. Karp, Reducibility among combinatorial problems, in *Complexity of Computer Computations.* Springer, Berlin/Heidelberg, Germany pp. 85–103 (1972).

[72] M. Kubale, *Graph Colorings*, Vol. 352. American Mathematical Society, Providence, RI, USA (2004).

[73] K. Appel and W. Haken, Every planar map is four colorable, *Bulletin of the American Mathematical Society* **82**(5), pp. 711–712 (1976).

[74] G. B. Kolata, The four-color conjecture: A computer-aided proof, *Science* **193**(4253), pp. 564–565 (1976).

[75] G. D. Birkhoff, The reducibility of maps, *American Journal of Mathematics* **35**(2), pp. 115–128 (1913).

[76] G. D. Birkhoff and D. C. Lewis, Chromatic polynomials, *Transactions of the American Mathematical Society* **60**(3), pp. 355–451 (1946).

[77] H. Whitney, The coloring of graphs, *Annals of Mathematics*, **33**(4), pp. 688–718 (1932).

[78] W. Tutte, Graph-polynomials, *Advances in Applied Mathematics* **32**(1–2), pp. 5–9 (2004).

[79] A. V. Aho, J. E. Hopcroft and J. D. Ullman, *The Design and Analysis of Computer Algorithms.* Addison-Wesley, Boston, MA, USA (1974).

[80] J. B. Fraleigh, *A First Course in Abstract Algebra*, 6th edn. Addison-Wesley, Boston, MA, USA (1999).

[81] A. Lubiw, Some NP-complete problems similar to graph isomorphism, *SIAM Journal on Computing* **10**(1), pp. 11–21 (1981).
[82] G. Di Battista, R. Tamassia and I. G. Tollis, Area requirement and symmetry display of planar upward drawings. *Discrete & Computational Geometry* **7**(4), pp. 381–401 (1992).
[83] S.-H. Hong, Drawing graphs symmetrically in three dimensions, in *International Symposium on Graph Drawing*. Springer, Berlin/Heidelberg, Germany, pp. 189–204 (2001).
[84] D. König, *Theorie der endlichen und unendlichen graphen*. Akad, Verlagsgeselschaft, Leipzig (1936).
[85] R. Frucht, Herstellung von graphen mit vorgegebener abstrakter gruppe, *Compositio Mathematica* **6**, pp. 239–250 (1939).
[86] J. J. Sylvester, XLVII. Additions to the articles in the september number of this journal, "on a new class of theorems," and on Pascal's theorem, *The London, Edinburgh, and Dublin Philosophical Magazine and Journal of Science* **37**(251), pp. 363–370 (1850).
[87] J. Dossey, A. Otto, L. Spence and C. V. Eynden, *Discrete Mathematics*. Pearson, Boston, MA, USA (2017).
[88] A. Cayley, II. A memoir on the theory of matrices, *Philosophical Transactions of the Royal Society of London*, **148**, pp. 17–37 (1858).
[89] C. E. Cullis, *Matrices and Determinoids*, Vol. 1. Cambridge University Press, Cambridge, England (1913).
[90] G. Pólya and G. Szegő, *Aufgaben und Lehrsätze aus der Analysis*, Vol. 1. Springer, Berlin/Heidelberg, Germany (1925).
[91] F. Harary, The determinant of the adjacency matrix of a graph, *SIAM Review* **4**(3), pp. 202–210 (1962).
[92] L. Von Collatz and U. Sinogowitz, Spektren endlicher grafen, in *Abhandlungen aus dem Mathematischen Seminar der Universität Hamburg*, Vol. 21. Springer, Berlin/Heidelberg, Germany, pp. 63–77 (1957).
[93] F. Harary, A graph theoretic method for the complete reduction of a matrix with a view toward finding its eigenvalues, *Journal of Mathematics and Physics* **38**(1–4), pp. 104–111 (1959).
[94] S. Bowles, Technical change and the profit rate: A simple proof of the Okishio theorem, *Cambridge Journal of Economics* **5**(2), pp. 183–186 (1981).
[95] E. Landau, Zur relativen wertbemessung der turnierresultate, *Deutsches Wochenschach* **11**, pp. 366–369 (1895).
[96] E. G. H. Landau, *Über preisverteilung bei spielturnieren*. Zeitschrift für Mathematik und Physik, **63**, pp. 192–202 (1914).
[97] C. D. Meyer, *Matrix Analysis and Applied Linear Algebra*. SIAM Publishing, Philadelphia, PA, USA (2001).

[98] L. Spizzirri, Justification and application of eigenvector centrality (2011), http://www.math.washington.edu/~morrow/336_11/papers/leo.pdf (Last Checked: July 20, 2011).

[99] S. Ross, *Introduction to Probability Models.* Academic Press, Cambridge, MA, USA (2010).

[100] B. N. Datta, *Numerical Linear Algebra.* Brooks/Cole, Boston, MA, USA (1995).

[101] S. Lang, *Linear Algebra.* Springer-Verlag, Berlin/Heidelberg, Germany (1987).

[102] M. Fiedler, Algebraic connectivity of graphs, *Czechoslovak Mathematical Journal* **23**(98), pp. 298–305 (1973).

[103] S. Vigna, Spectral ranking, *Network Science* **4**(4), pp. 433–445 (2016).

[104] J. R. Seeley, The net of reciprocal influence. A problem in treating sociometric data, *Canadian Journal of Experimental Psychology* **3**, p. 234 (1949).

[105] T.-H. Wei, *Algebraic Foundations of Ranking Theory*, Ph.D. thesis, University of Cambridge (1952).

[106] C. Berge, *Théorie des graphes.* Paris, France, Wiley, Hoboken, NJ, USA (1958).

[107] L. Page, S. Brin, R. Motwani and T. Winograd, The pagerank citation ranking: Bringing order to the web. Technical Report, Stanford InfoLab (1999).

[108] K. Bryan and T. Leise, The $25,000,000,000 eigenvector: The linear algebra behind google, *SIAM Review* **48**(3), pp. 569–581 (2006).

[109] D. Austin, How google finds your needle in the web's haystack, *American Mathematical Society Feature Column* **10**(12) (2006).

[110] L. Rabiner and B. Juang, An introduction to hidden Markov models, *IEEE ASSP Magazine* **3**(1), pp. 4–16 (1986).

[111] A. Ng, M. Jordan and Y. Weiss, On spectral clustering: Analysis and an algorithm, *Advances in Neural Information Processing Systems* **14** (2001).

[112] M. Newman and M. Girvan, Finding and evaluating community structure in networks, *Physical Review E* **69**, p. 026113 (2004).

[113] K. D. Cole, A. Riensche and P. K. Rao, Discrete Green's functions and spectral graph theory for computationally efficient thermal modeling, *International Journal of Heat and Mass Transfer* **183**, p. 122112 (2022).

[114] S. Motsch and E. Tadmor, Heterophilious dynamics enhances consensus, *SIAM Review* **56**(4), pp. 577–621 (2014).

[115] L. A. Wolsey and G. L. Nemhauser, *Integer and Combinatorial Optimization.* Wiley-Interscience, Hoboken, NJ, USA (1999).

[116] G. Sierksma and Y. Zwols, *Linear and Integer Optimization: Theory and Practice*. CRC Press, Boca Raton, FL, USA (2015).
[117] D. J. Albers and C. Reid, An interview with George B. Dantzig: The father of linear programming, *The College Mathematics Journal* **17**(4), pp. 292–314 (1986).
[118] R. W. Cottle, George B. Dantzig: Operations research icon, *Operations Research* **53**(6), pp. 892–898 (2005).
[119] R. W. Cottle, George B. Dantzig: A legendary life in mathematical programming, *Mathematical Programming* **105**(1), pp. 1–8 (2006).
[120] G. B. Dantzig, Reminiscences about the origins of linear programming, in *Mathematical Programming the State of the Art*. Springer, Berlin/Heidelberg, Germany, pp. 78–86 (1983).
[121] L. G. Khachiyan, A polynomial algorithm in linear programming, *Doklady Akademii Nauk* (Russian Academy of Sciences) **244**, pp. 1093–1096 (1979).
[122] N. Karmarkar, A new polynomial-time algorithm for linear programming, in *Proceedings of the Sixteenth Annual ACM Symposium on Theory of Computing*, pp. 302–311 (1984).
[123] D. A. Spielman and S.-H. Teng, Smoothed analysis of algorithms: Why the simplex algorithm usually takes polynomial time, *Journal of the ACM (JACM)* **51**(3), pp. 385–463 (2004).
[124] J. Orlin, A faster strongly polynomial minimum cost flow algorithm, in *Proceedings of the Twentieth Annual ACM Symposium on Theory of Computing*, pp. 377–387 (1988).
[125] D. R. Fulkerson, An out-of-kilter method for minimal-cost flow problems, *Journal of the Society for Industrial and Applied Mathematics* **9**(1), pp. 18–27 (1961).
[126] A. V. Goldberg and R. E. Tarjan, A new approach to the maximum-flow problem, *Journal of the ACM (JACM)* **35**(4), pp. 921–940 (1988).
[127] T. C. Koopmans, Optimum utilization of the transportation system, *Econometrica: Journal of the Econometric Society*, **17**, pp. 136–146 (1949).
[128] J. B. Orlin, S. A. Plotkin and É. Tardos, Polynomial dual network simplex algorithms, *Mathematical Programming* **60**(1), pp. 255–276 (1993).
[129] J. B. Orlin, A polynomial time primal network simplex algorithm for minimum cost flows, *Mathematical Programming* **78**(2), pp. 109–129 (1997).
[130] R. E. Tarjan, Dynamic trees as search trees via euler tours, applied to the network simplex algorithm, *Mathematical Programming* **78**(2), pp. 169–177 (1997).

[131] J. Edmonds, Paths, trees, and flowers, *Canadian Journal of Mathematics* **17**, pp. 449–467 (1965a).

[132] J. Edmonds, Maximum matching and a polyhedron with 0, 1-vertices, *Journal of Research of the National Bureau of Standards B* **69**(125–130), pp. 55–56 (1965b).

[133] A. Schrijver *et al.*, *Combinatorial Optimization: Polyhedra and Efficiency*, Vol. 24. Springer, Berlin/Heidelberg, Germany (2003).

[134] C. H. Papadimitriou and K. Steiglitz, *Combinatorial Optimization: Algorithms and Complexity*. Dover Press, Mineola, NY, USA (1998).

[135] J. W. Demmel, *Applied Numerical Linear Algebra*. SIAM Publishing, Philadelphia, PA, USA (1997).

[136] G. Strang, *Linear Algebra and Its Applications*. Brooks/Cole, Boston, MA, USA (2005).

Index

D

E

W

Z

M